VOLCANIC ERUPTIONS AND THEIR REPOSE, UNREST, PRECURSORS, AND TIMING

Committee on Improving Understanding of Volcanic Eruptions

Committee on Seismology and Geodynamics

Board on Earth Sciences and Resources

Division on Earth and Life Studies

A Consensus Study Report of

The National Academies of

SCIENCES · ENGINEERING · MEDICINE

THE NATIONAL ACADEMIES PRESS

Washington, DC

www.nap.edu

THE NATIONAL ACADEMIES PRESS • 500 Fifth Street, NW • Washington, DC 20001

This activity is supported by the National Academies of Sciences, Engineering, and Medicine's Day Fund, the National Aeronautics and Space Administration under Grant No. NNX15AT28G, the National Science Foundation under Grant No. EAR-1547098, and the U.S. Geological Survey under Grant No. G15AC00348. Any opinions, findings, conclusions, or recommendations expressed in this publication do not necessarily reflect the views of any organization or agency that provided support for the project.

International Standard Book Number-13: 978-0-309-45412-4
International Standard Book Number-10: 0-309-45412-3
Digital Object Identifier: https://doi.org/10.17226/24650

Additional copies of this publication are available for sale from the National Academies Press, 500 Fifth Street, NW, Keck 360, Washington, DC 20001; (800) 624-6242 or (202) 334-3313; http://www.nap.edu.

Printed in the United States of America

Cover: A grand challenge in the report is to understand the full life cycle of volcanoes. The life cycle consists of the repose period; the unrest, which is only sometimes followed by eruption; the precursors (geophysical or geochemical changes), which are followed by eruption; and the eruption itself. A life cycle involves a set of events at various times, so it is important to quantify the timing of unrest, precursors, and the creation of magma bodies as well as the time scales for eruptive processes, including the initiation, duration, and ending of an eruption. Cover image: Augustine volcano, Alaska, on March 27, 2006. Courtesy of photographer Cyrus Read and the Alaska Volcano Observatory, U.S. Geological Survey (USGS). Cover designed by Francesca Moghari.

Chapter opening photos: (Chapter 1) Active Caliente Vent of Santiaguito, Guatemala, in 2007. Courtesy of Jeffrey Johnson, Boise State University. (Chapter 2) Lava lake spattering at the summit of Kilauea volcano, Hawaii, in January 2017. Courtesy of the USGS Hawaiian Volcano Observatory. (Chapter 3) Monitoring and news trucks at Mount St. Helens in 2004. Courtesy of Larry Mastin, USGS. (Chapter 4) Geothermal geysers, El Tatio, Chile. Courtesy of Michael Manga, University of California, Berkeley. (Chapter 5) Obsidian dome in Long Valley Caldera, California. Courtesy of Michael Manga, University of California, Berkeley. (Chapter 6) Lava stream from the lava tube at Kilauea into the ocean. Courtesy of the USGS Hawaiian Volcano Observatory.

Suggested citation: National Academies of Sciences, Engineering, and Medicine. 2017. *Volcanic Eruptions and Their Repose, Unrest, Precursors, and Timing.* Washington, DC: The National Academies Press. doi: https://doi.org/10.17226/24650.

The National Academies of

SCIENCES • ENGINEERING • MEDICINE

The **National Academy of Sciences** was established in 1863 by an Act of Congress, signed by President Lincoln, as a private, nongovernmental institution to advise the nation on issues related to science and technology. Members are elected by their peers for outstanding contributions to research. Dr. Marcia McNutt is president.

The **National Academy of Engineering** was established in 1964 under the charter of the National Academy of Sciences to bring the practices of engineering to advising the nation. Members are elected by their peers for extraordinary contributions to engineering. Dr. C. D. Mote, Jr., is president.

The **National Academy of Medicine** (formerly the Institute of Medicine) was established in 1970 under the charter of the National Academy of Sciences to advise the nation on medical and health issues. Members are elected by their peers for distinguished contributions to medicine and health. Dr. Victor J. Dzau is president.

The three Academies work together as the **National Academies of Sciences, Engineering, and Medicine** to provide independent, objective analysis and advice to the nation and conduct other activities to solve complex problems and inform public policy decisions. The National Academies also encourage education and research, recognize outstanding contributions to knowledge, and increase public understanding in matters of science, engineering, and medicine.

Learn more about the National Academies of Sciences, Engineering, and Medicine at **www.nationalacademies.org**.

The National Academies of

SCIENCES • ENGINEERING • MEDICINE

Consensus Study Reports published by the National Academies of Sciences, Engineering, and Medicine document the evidence-based consensus on the study's statement of task by an authoring committee of experts. Reports typically include findings, conclusions, and recommendations based on information gathered by the committee and the committee's deliberations. Each report has been subjected to a rigorous and independent peer-review process and it represents the position of the National Academies on the statement of task.

Proceedings published by the National Academies of Sciences, Engineering, and Medicine chronicle the presentations and discussions at a workshop, symposium, or other event convened by the National Academies. The statements and opinions contained in proceedings are those of the participants and are not endorsed by other participants, the planning committee, or the National Academies.

For information about other products and activities of the National Academies, please visit www.nationalacademies.org/about/whatwedo.

COMMITTEE ON IMPROVING UNDERSTANDING OF VOLCANIC ERUPTIONS

MICHAEL MANGA, *Chair*, University of California, Berkeley
SIMON A. CARN, Michigan Technological University, Houghton
KATHARINE V. CASHMAN, NAS,[1] University of Bristol, United Kingdom
AMANDA B. CLARKE, Arizona State University, Tempe
CHARLES B. CONNOR, University of South Florida, Tampa
KARI M. COOPER, University of California, Davis
TOBIAS FISCHER, University of New Mexico, Albuquerque
BRUCE HOUGHTON, University of Hawaii at Manoa
JEFFREY B. JOHNSON, Boise State University, Idaho
TERRY A. PLANK, NAS, Columbia University, New York
DIANA C. ROMAN, Carnegie Institution for Science, Washington, DC
PAUL SEGALL, NAS, Stanford University, California

Committee on Seismology and Geodynamics Liaison

STEPHEN McNUTT, University of South Florida, Tampa

National Academies of Sciences, Engineering, and Medicine Staff

ANNE M. LINN, Scholar
NICHOLAS ROGERS, Financial and Research Associate
COURTNEY R. GIBBS, Administrative Coordinator

[1] National Academy of Sciences.

COMMITTEE ON SEISMOLOGY AND GEODYNAMICS

RICHARD M. ALLEN, *Chair*, University of California, Berkeley
THORSTEN W. BECKER, The University of Texas at Austin
CYNTHIA EBINGER, Tulane University, New Orleans, Louisiana
STEVEN JACOBSEN, Northwestern University, Evanston, Illinois
LISA G. LUDWIG, University of California, Irvine
STEPHEN McNUTT, University of South Florida, Tampa
MATTHEW PRITCHARD, Cornell University, Ithaca, New York
MAYA TOLSTOY, Lamont-Doherty Earth Observatory, Palisades, New York
JEROEN TROMP, Princeton University, New Jersey
WILLIAM WALTER, Lawrence Livermore National Laboratory, Livermore, California
SHERILYN WILLIAMS-STROUD, Confractus, Inc., Pasadena, California

National Academies of Sciences, Engineering, and Medicine Staff

DEBORAH GLICKSON, Senior Program Officer
COURTNEY R. GIBBS, Administrative Coordinator

BOARD ON EARTH SCIENCES AND RESOURCES

GENE WHITNEY, *Chair*, Congressional Research Service (Retired), Washington, DC
R. LYNDON (LYN) ARSCOTT, International Association of Oil & Gas Producers (Retired), Danville, California
CHRISTOPHER (SCOTT) CAMERON, GeoLogical Consulting, LLC, Houston, Texas
RODNEY C. EWING, NAE,[1] Freeman Spogli Institute for International Studies and Stanford University, California
CAROL P. HARDEN, The University of Tennessee, Knoxville
T. MARK HARRISON, University of California, Los Angeles
THORNE LAY, NAS,[2] University of California, Santa Cruz
ANN S. MAEST, Buka Environmental, Boulder, Colorado
ZELMA MAINE-JACKSON, Washington State Department of Ecology, Nuclear Waste Program, Richland
MARTIN W. McCANN, Jack R. Benjamin and Associates and Stanford University, Menlo Park, California
JAMES M. ROBERTSON, Wisconsin Geological and Natural History Survey, Madison
JAMES SLUTZ, National Petroleum Council, Washington, DC
SHAOWEN WANG, University of Illinois at Urbana-Champaign

National Academies of Sciences, Engineering, and Medicine Staff

ELIZABETH A. EIDE, Senior Board Director
ANNE M. LINN, Scholar
DEBORAH GLICKSON, Senior Program Officer
SAMMANTHA L. MAGSINO, Senior Program Officer
NICHOLAS D. ROGERS, Financial and Research Associate
COURTNEY R. GIBBS, Administrative Coordinator
ERIC J. EDKIN, Senior Program Assistant
RAYMOND M. CHAPPETTA, Senior Program Assistant

[1] National Academy of Engineering.
[2] National Academy of Sciences.

Acknowledgments

This Consensus Study Report was reviewed in draft form by individuals chosen for their diverse perspectives and technical expertise. The purpose of this independent review is to provide candid and critical comments that will assist the National Academies of Sciences, Engineering, and Medicine in making each published report as sound as possible and to ensure that it meets the institutional standards for quality, objectivity, evidence, and responsiveness to the study charge. The review comments and draft manuscript remain confidential to protect the integrity of the deliberative process.

We thank the following individuals for their review of this report:

Emily E. Brodsky, University of California, Santa Cruz
Donald J. DePaolo, Lawrence Berkeley National Laboratory, Berkeley, California
Josef D. Dufek, Georgia Institute of Technology, Atlanta
Marie Edmonds, University of Cambridge, United Kingdom
Guido Giordano, Roma Tre University, Italy
Warner Marzocchi, Istituto Nazionale di Geofisica e Vulcanologia, Rome, Italy
J.R. Anthony Pearson, Schlumberger Cambridge Research, United Kingdom
Matthew Pritchard, Cornell University, Ithaca, New York
Stephen Self, Nuclear Regulatory Commission, Alameda, California
S. Adam Soule, Woods Hole Oceanographic Institution, Massachusetts
Robert Wright, University of Hawaii at Manoa

Although the reviewers listed above provided many constructive comments and suggestions, they were not asked to endorse the conclusions or recommendations of this report nor did they see the final draft before its release. The review of this report was overseen by Robin K. McGuire, Lettis Consultants International, Inc., and E. Bruce Watson, Rensselaer Polytechnic Institute. They were responsible for making certain that an independent examination of this report was carried out in accordance with the standards of the National Academies and that all review comments were carefully considered. Responsibility for the final content rests entirely with the authoring committee and the National Academies.

The committee would like to thank the following individuals who shared their expertise with the committee through presentations, videos, or discussions: Amy Chen, Sonia Esperança, Dennis Geist, Jake Lowenstern, Charles Mandeville, Seth Moran, Tina Neal, John Pallister, Benjamin Phillips, James Rustad, and Jennifer Wade. Particular thanks go to the participants in an international workshop (see Appendix B), who often came long distances and kept to a demanding schedule of presentations and working group discussions. The committee is grateful for their contributions to the workshop and for the many informal discussions. The committee thanks Stephen McNutt, who brought valuable insights from the Committee on Seismology and Geodynamics. Finally, the committee thanks Anne Linn and Nicholas Rogers for their guidance and patience throughout the study.

Contents

Overview

Volcanic eruptions are common, with more than 50 volcanic eruptions in the United States alone in the past 31 years. These eruptions can have devastating economic and social consequences, even at great distances from the volcano. Fortunately many eruptions are preceded by unrest that can be detected using ground, airborne, and spaceborne instruments. Data from these instruments, combined with basic understanding of how volcanoes work, form the basis for forecasting eruptions—where, when, how big, how long, and the consequences. At the same time, monitoring data provide key insights into how volcanoes work.

We broadly understand why and where volcanoes exist, how the magma feeding the volcano is generated and evolves, and how magma that erupts is distributed over Earth's surface. Yet our understanding is incomplete. What controls whether magma will erupt? What processes initiate eruptions? How quickly does magma rise to the surface? Which types of unrest are precursors to eruption rather than a return to dormancy? Which volcanoes are most likely to erupt in the coming decades?

Major improvements in understanding and forecasting are possible through enhanced monitoring combined with advances in experimental and mathematical models for volcanic processes. In the United States, fewer than half of the 169 potentially active volcanoes have even one seismometer to detect the small earthquakes that signal underground magma movement. Only three have continuous gas measurements—gas matters because it drives eruptions.

The title of this report reflects one of the grand challenges in volcano science: to document and understand the repose, unrest, precursors, and timing of eruptions during the entire life cycle of volcanoes. At present, our understanding is biased because the necessary observations are available for only a few volcanoes. Moreover, activity at these volcanoes represents only a small fraction of the diversity of eruptions on Earth.

A lack of monitoring hampers forecasting because most eruption forecasts are based on recognizing patterns in data. Models of volcanic processes provide a basis for closing observational gaps and hence could help improve forecasting. A second grand challenge is to develop quantitative models for the processes that govern volcanic eruptions and to use these models to forecast the size, duration, and hazard of eruptions.

A third grand challenge is to develop a coordinated community of scientists who will make this happen. Foremost this requires effective integration of the complementary research and monitoring roles of universities, the U.S. Geological Survey, and other government agencies. In addition, volcano science draws on a large number of disciplines (e.g., geology, geophysics, geochemistry) and approaches (e.g., remote sensing, high-performance computing), and vehicles are needed to support interdisciplinary research and training, including community collaborations and education at all levels.

Although these grand challenges are large in scope and require great effort, achieving them would yield new understanding of how volcanoes work and their consequences, and improve volcano eruption planning and warning for all of society.

Summary

Volcanoes are a key part of the Earth system, and open a window into the inner workings of the planet. More than a dozen volcanoes are usually erupting on Earth at any given time. Some of these eruptions are devastating, killing people, damaging homes and infrastructure, altering landscapes, and even disrupting climate. Fortunately, many eruptions are preceded by signs of unrest (precursors) that can be used to anticipate eruptions and support disaster planning.

Accurate forecasts of the likelihood and magnitude of an eruption in a specified timeframe are rooted in a scientific understanding of the processes that govern the storage, ascent, and eruption of magma. Yet our understanding of volcanic systems is incomplete and biased by the limited number of volcanoes and eruption styles observed with advanced instrumentation. Eruption behaviors are diverse (e.g., violently explosive or gently effusive, intermittent or sustained, last hours or decades) and may change over time at a volcano. More accurate and societally useful forecasts of eruptions and their hazards are possible by using new observations and models of volcanic processes.

At the request of managers at the National Aeronautics and Space Administration, the National Science Foundation (NSF), and the U.S. Geological Survey (USGS), the National Academies of Sciences, Engineering, and Medicine established a committee to undertake the following tasks:

- Summarize current understanding of how magma is stored, ascends, and erupts.
- Discuss new disciplinary and interdisciplinary research on volcanic processes and precursors that could lead to forecasts of the type, size, and timing of volcanic eruptions.
- Describe new observations or instrument deployment strategies that could improve quantification of volcanic eruption processes and precursors.
- Identify priority research and observations needed to improve understanding of volcanic eruptions and to inform monitoring and early warning efforts.

These four tasks are closely related. Improved understanding of volcanic processes guides monitoring efforts and improves forecasts. In turn, improved monitoring provides the insights and constraints to better understand volcanic processes. This report identifies key science questions, research and observation priorities, and approaches for building a volcano science community capable of tackling them. The discussion below first summarizes common themes among these science questions and priorities, and then describes ambitious goals (grand challenges) for making major advances in volcano science.

KEY QUESTIONS AND RESEARCH AND OBSERVATION PRIORITIES

Many fundamental aspects of volcanoes are understood conceptually and often quantitatively. Plate tectonics and mantle convection explain where volcanoes occur. We understand how magma is initially created in Earth's mantle, how it rises toward the surface, that it can be stored and evolve in magma chambers within the crust, and that a number of processes initiate eruptions. We understand in general terms why some magmas erupt explosively and others do not, and why some volcanoes erupt more often than others. High-resolution observations and models combined provide a detailed and quantitative picture of eruptions once they begin.

Our understanding is incomplete, however, especially those aspects of volcano behavior that define the timing, duration, style, size, and consequences of eruptions. Additional questions relate to our ability to forecast eruptions. What processes produce commonly observed geophysical and geochemical precursors? What factors determine if and when unrest will be followed by eruption? How rapidly do magmas mobilize prior to eruption? Which volcanoes are most likely to erupt in coming years and decades? And we are only beginning to decipher the impacts of large volcanic eruptions on Earth's climate and biosphere.

Our understanding of the entire life cycle and diversity of volcanoes—from their conception in the mantle to their periods of repose, unrest, and eruption to their eventual demise—is poised for major advances over the next decades. Exciting advances in our ability to observe volcanoes—including satellite measurements of ground deformation and gas emissions, drone observations, advanced seismic monitoring, and real-time, high-speed acquisition of data during eruptions—await broad application to volcanic systems. Parallel advances in analytical capabilities to decipher the history of magmas, and in conceptual, experimental, and numerical models of magmatic and volcanic phenomena, both below and above ground, will provide new insights on the processes that govern the generation and eruption of magma and greatly improve the quality of short-term, months to minutes, forecasts. The time is ripe to test these models with observations from new instrumentation, data collected on fine temporal and spatial scales, and multidisciplinary synthesis.

Four common themes emerged from the research priorities detailed in the following chapters:

1. Develop multiscale models that capture critical processes, feedbacks, and thresholds to advance understanding of volcanic processes and the consequences of eruptions on Earth systems.

Advances will come from measurements of physical and chemical properties of magmas and erupted materials, deciphering the history of magmas (before and during eruption) recorded in their crystals and bubbles, and developing new models that account for the numerous interacting processes and vast range of scales, from microscopic ash particles and crystals, to eruption columns that extend to the stratosphere.

2. Collect high-resolution measurements at more volcanoes and throughout their life cycle to overcome observational bias.

Few volcanoes have a long record of monitoring data. New and expanded networks of ground, submarine, airborne, and satellite sensors that characterize deformation, gases, and fluids are needed to document volcanic processes during decade-long periods of repose and unrest. High-rate, near-real-time measurements are needed to capture eruptions as they occur, and efficient dissemination of information is needed to formulate a response. Both rapid response and sustained monitoring are required to document the life cycle of volcanoes. Monitoring and understanding volcanic processes go hand-in-hand: Different types of volcanoes have different life c`ycles and behaviors, and hence merit different monitoring strategies.

3. Synthesize a broad range of observations, from the subsurface to space, to interpret unrest and forecast eruption size, style, and duration.

Physics-based models promise to improve forecasts by assimilating monitoring data and observations. Progress in forecasting also requires theoretical and experimental advances in understanding eruption processes, characterization of the thermal and mechanical properties of magmas and their host rocks, and model validation and verification. Critical to eruption forecasting is reproducing with models and documenting with measurements the emergent precursory phenomena in the run-up to eruption.

4. Obtain better chronologies and rates of volcanic processes.

Long-term forecasts rely on understanding the geologic record of eruptions preserved in volcanic deposits on land, in marine and lake sediments, and in ice cores. Secondary hazards that are not part of the eruption itself, such as mud flows and floods, need to be better studied, as they can have more devastating consequences than the eruption. Understanding the effects of eruptions on other Earth systems, including climate, the oceans, and landscapes, will take coordinated efforts across disciplines. Progress in long-term forecasts, years to decades, requires open-access databases that document the full life cycle of volcanoes.

GRAND CHALLENGES

The key science questions, research and observation priorities, and new approaches highlighted in this report can be summarized by three overarching grand challenges. These challenges are grand because they are large in scope and would substantially advance the field, and they are challenges because great effort will be needed. Figure S.1 illustrates these challenges using the example of the 2016 eruption of Pavlof volcano, Alaska. The volcanic hazards and eruption history of Pavlof are summarized by Waythomas et al. (2006).

1. Forecast the onset, size, duration, and hazard of eruptions by integrating observations with quantitative models of magma dynamics.

A principal goal of volcano science is to reduce the adverse impacts of volcanism on humanity, which requires accurate forecasts. Most current eruption forecasts use pattern recognition in monitoring and geologic data. Such approaches have led to notable forecasts in some cases, but their use is limited because volcanoes evolve over time, there is a great diversity of volcano behavior, and we have no experience with many of the potentially most dangerous volcanoes. A major challenge is to develop forecasting models based instead on physical and chemical processes, informed by monitoring. This approach is used in weather forecasting. Addressing this challenge requires an understanding of the basic processes of magma storage and ascent as well as thresholds of eruption initiation. This understanding and new discoveries will emerge from new observations, experimental measurements, and modeling approaches. Models are important because they capture our conceptual and quantitative understanding. Experiments test our understanding. Relating models to observations requires multiple types of complementary data collected over an extended period of time.

2. Quantify the life cycles of volcanoes globally and overcome our current biased understanding.

Determining the life cycle of volcanoes is key for interpreting precursors and unrest, revealing the processes that govern the initiation and duration of eruptions, and understanding how volcanoes evolve between eruptions. Our understanding is biased by an emphasis over the last few decades of observation with modern instruments, and most of these well-studied eruptions have been small events that may not scale to the largest and most devastating eruptions. Strategic deployment of instruments on volcanoes with different characteristics would help build the requisite knowledge and confidence to make useful forecasts. For every volcano in the United States, a realistic goal is to have at least one seismometer to record the small earthquakes that accompany magma movement. Even in the United States, less than half of potentially active volcanoes have a seismometer, and less than 2 percent have continuous gas measurements. Global and daily satellite images of deformation, and the ability to measure passive CO_2 degassing from space would fill critical observational gaps. Geologic and geophysical studies are required to extend understanding of the life cycle of volcanoes to longer periods of time. On shorter time scales, satellite measurements, emerging technologies such as drones, and expansion of ground-based monitoring networks promise to document processes that remain poorly understood.

3. Develop a coordinated volcano science community to maximize scientific returns from any volcanic event.

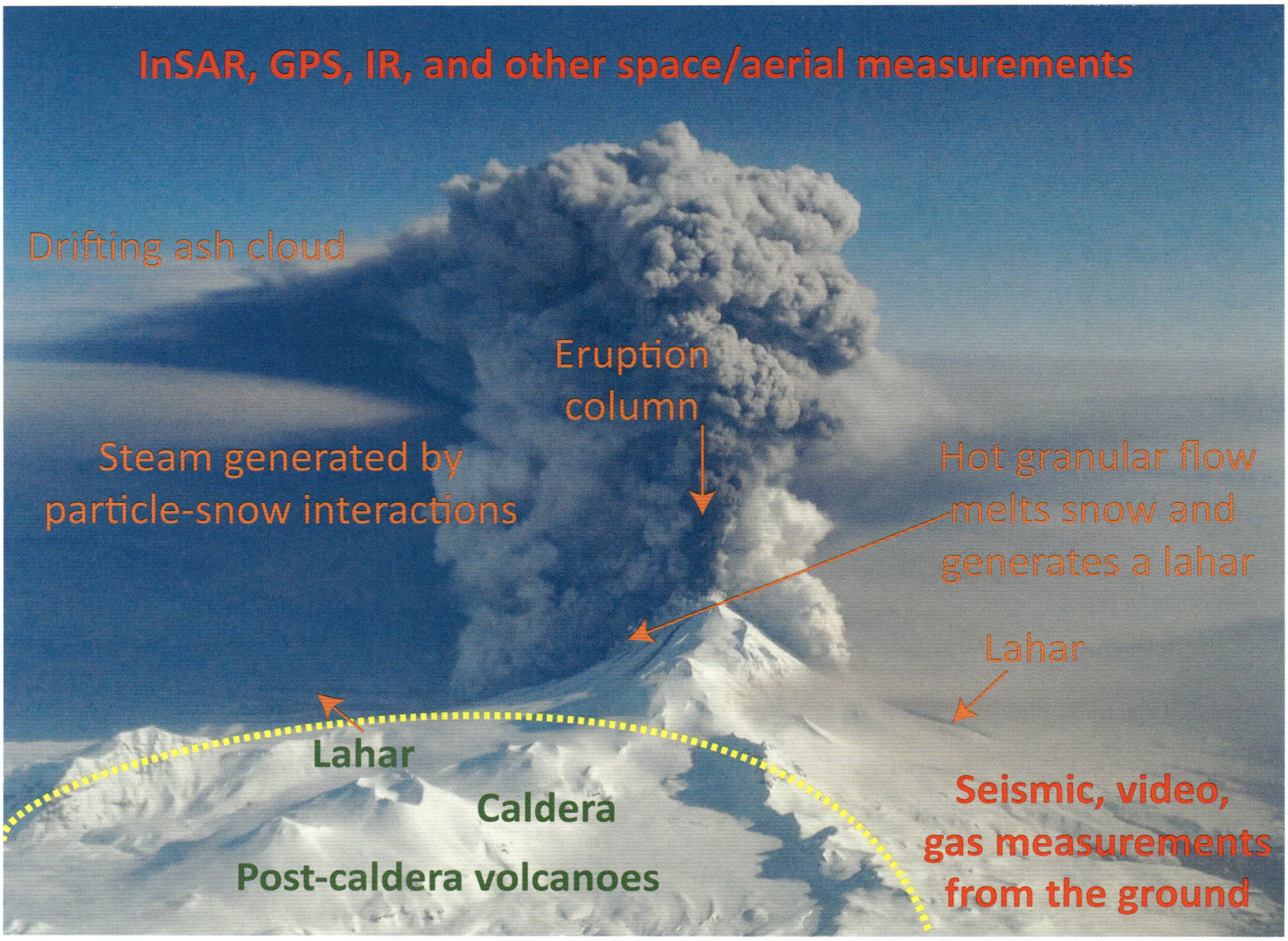

FIGURE S.1 Pavlof Volcano erupted on March 28, 2016, creating a drifting cloud of ash, pyroclastic density currents, and lahars. Pavlof is next to the Emmons Lake caldera, one of the largest calderas in Alaska. Words in orange identify processes for which quantitative models can be developed and integrated with observations (challenge 1). Features in green provide insight into the life cycle of volcanoes (challenge 2). Red words show the type of measurements that can be made to understand both the processes that govern volcanic eruptions and improve forecasting, and whose measurement and use require an effective and integrated volcano science community (challenge 3). NOTE: GPS, Global Positioning System; InSAR, interferometric synthetic aperture radar; IR, infrared. SOURCE: Background photo courtesy of Nahshon Almandmoss, U.S. Coast Guard.

The volcano science community needs to be prepared to capitalize on the data and insights gained from eruptions as they happen. This will come from effective integration of the complementary research and monitoring roles by universities, the USGS, and other government agencies. Volcano science is fundamentally interdisciplinary and the necessary expertise is spread across these institutions. The science is also international, because every volcano provides insights on processes that drive eruptions. Volcanic eruptions can have global impacts and so demand international collaboration and cooperation. New vehicles are needed to support interdisciplinary research and training, including community collaboration and education at all levels. Examples of similar successful programs in other fields include NSF's Cooperative Studies of the Earth's Deep Interior program for interdisciplinary research and National Earthquake Hazards Reduction

Program for federal government agency–academic partnerships.

Results of the above investments in science will be most evident to the public in improved planning and warning and, ideally, a deeper appreciation of this amazing natural phenomenon.

1

Introduction

Volcanoes are a key part of the Earth system. Most of Earth's atmosphere, water, and crust were delivered by volcanoes, and volcanoes continue to recycle earth materials. Volcanic eruptions are common. More than a dozen are usually erupting at any time somewhere on Earth, and close to 100 erupt in any year (Loughlin et al., 2015).

Volcano landforms and eruptive behavior are diverse, reflecting the large number and complexity of interacting processes that govern the generation, storage, ascent, and eruption of magmas. Eruptions are influenced by the tectonic setting, the properties of Earth's crust, and the history of the volcano. Yet, despite the great variability in the ways volcanoes erupt, eruptions are all governed by a common set of physical and chemical processes. Understanding how volcanoes form, how they erupt, and their consequences requires an understanding of the processes that cause rocks to melt and change composition, how magma is stored in the crust and then rises to the surface, and the interaction of magma with its surroundings. Our understanding of how volcanoes work and their consequences is also shared with the millions of people who visit U.S. volcano national parks each year.

Volcanoes have enormous destructive power. Eruptions can change weather patterns, disrupt climate, and cause widespread human suffering and, in the past, mass extinctions. Globally, volcanic eruptions caused about 80,000 deaths during the 20th century (Sigurdsson et al., 2015). Even modest eruptions, such as the 2010 Eyjafjallajökull eruption in Iceland, have multibillion-dollar global impacts through disruption of air traffic. The 2014 steam explosion at Mount Ontake, Japan, killed 57 people without any magma reaching the surface. Many volcanoes in the United States have the potential for much larger eruptions, such as the 1912 eruption of Katmai, Alaska, the largest volcanic eruption of the 20th century (Hildreth and Fierstein, 2012). The 2008 eruption of the unmonitored Kasatochi volcano, Alaska, distributed volcanic gases over most of the continental United States within a week (Figure 1.1).

Finally, volcanoes are important economically. Volcanic heat provides low-carbon geothermal energy. U.S. generation of geothermal energy accounts for nearly one-quarter of the global capacity (Bertani, 2015). In addition, volcanoes act as magmatic and hydrothermal distilleries that create ore deposits, including gold and copper ores.

Moderate to large volcanic eruptions are infrequent yet high-consequence events. The impact of the largest possible eruption, similar to the super-eruptions at Yellowstone, Wyoming; Long Valley, California; or Valles Caldera, New Mexico, would exceed that of any other terrestrial natural event. Volcanoes pose the greatest natural hazard over time scales of several decades and longer, and at longer time scales they have the potential for global catastrophe (Figure 1.2). While

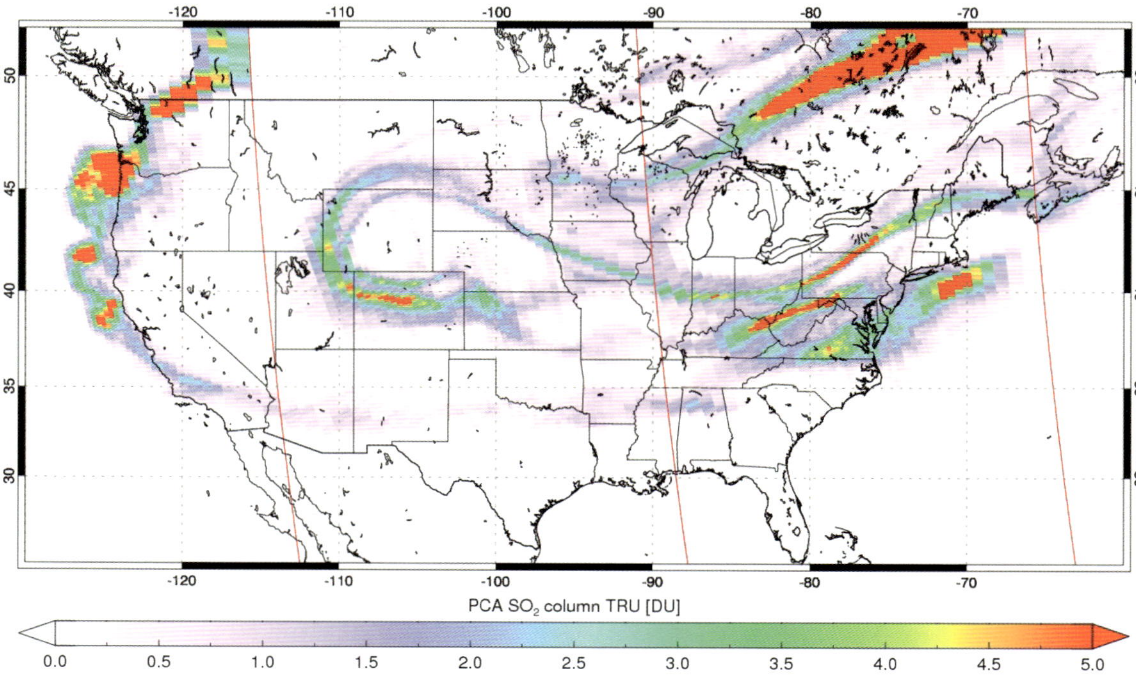

FIGURE 1.1 NASA Ozone Monitoring Instrument observations of the SO_2 cloud produced by the August 7, 2008, eruption of Kasatochi (Aleutian Islands, Alaska) drifting over the lower 48 states and Canada on August 15, 2008. Satellite observations such as these are crucial for mitigating aviation hazards due to drifting volcanic clouds and for assessing the impact of volcanic eruptions on Earth's atmosphere and climate. SOURCE: Adapted from Krotkov et al. (2010).

the continental United States has not suffered a fatal eruption since 1980 at Mount St. Helens, the threat has only increased as more people move into volcanic areas.

Volcanic eruptions evolve over very different temporal and spatial scales than most other natural hazards (Figure 1.3). In particular, many eruptions are preceded by signs of unrest that can serve as warnings, and an eruption itself often persists for an extended period of time. For example, the eruption of Kilauea Volcano in Hawaii has continued since 1983. We also know the locations of many volcanoes and, hence, where most eruptions will occur. For these reasons, the impacts of at least some types of volcanic eruptions should be easier to mitigate than other natural hazards.

Anticipating the largest volcanic eruptions is possible. Magma must rise to Earth's surface and this movement is usually accompanied by precursors—changes in seismic, deformation, and geochemical signals that can be recorded by ground-based and space-borne instruments. However, depending on the monitoring infrastructure, precursors may present themselves over time scales that range from a few hours (e.g., 2002 Reventador, Ecuador, and 2015 Calbuco, Chile) to decades before eruption (e.g., 1994 Rabaul, Papua New Guinea). Moreover, not all signals of volcanic unrest are immediate precursors to surface eruptions (e.g., currently Long Valley, California, and Campi Flegrei, Italy).

Probabilistic forecasts account for this uncertainty using all potential eruption scenarios and all relevant data. An important consideration is that the historical record is short and biased. The instrumented record is even shorter and, for most volcanoes, spans only the last few decades—a miniscule fraction of their lifetime. Knowledge can be extended qualitatively using field studies of volcanic deposits, historical accounts, and proxy data, such as ice and marine sediment cores and speleothem (cave) records. Yet, these too are biased because they commonly do not record small to moderate eruptions.

Understanding volcanic eruptions requires contributions from a wide range of disciplines and approaches. Geologic studies play a critical role in reconstructing the past eruption history of volcanoes,

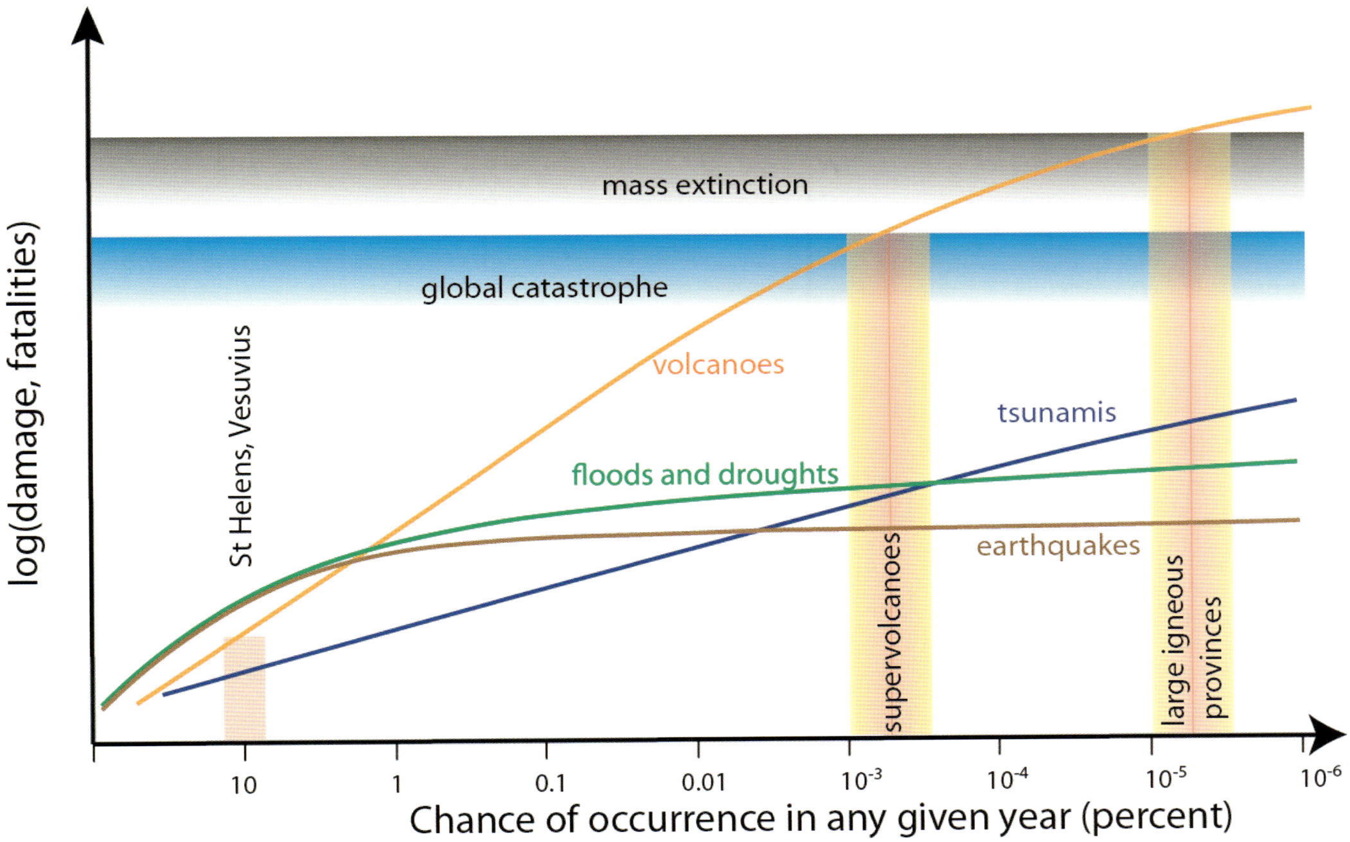

FIGURE 1.2 Qualitative comparison of consequences of selected natural hazards. Also shown are the frequency of events with magnitudes similar to Mount St. Helens (1980) and Vesuvius (79 AD), super-eruptions, and large igneous province eruptions. An exceptionally rare but very large supervolcano and large igneous province eruptions would have global consequences. In contrast, the maximum size of earthquakes limits their impacts. Tsunamis can be generated by earthquakes, landslides, volcanic eruptions, and asteroid impacts. The slope of the curves, while qualitative, reflects the relationship between event size and probability of occurrence: Earthquakes, and to a lesser extent floods and drought, saturate at a maximum size. SOURCE: Adapted from Plag et al. (2015).

especially of the largest events, and in regions with no historical or directly observed eruptions. Geochemical and geophysical techniques are used to study volcano processes at scales ranging from crystals to plumes of volcanic ash. Models reveal essential processes that control volcanic eruptions, and guide data collection. Monitoring provides a wealth of information about the life cycle of volcanoes and vital clues about what kind of eruption is likely and when it may occur.

1.1 OVERVIEW OF THIS REPORT

At the request of managers at the National Aeronautics and Space Administration (NASA), the National Science Foundation, and the U.S. Geological Survey (USGS), the National Academies of Sciences, Engineering, and Medicine established a committee to undertake the following tasks:

- Summarize current understanding of how magma is stored, ascends, and erupts.
- Discuss new disciplinary and interdisciplinary research on volcanic processes and precursors that could lead to forecasts of the type, size, and timing of volcanic eruptions.
- Describe new observations or instrument deployment strategies that could improve quantification of volcanic eruption processes and precursors.
- Identify priority research and observations needed to improve understanding of volcanic eruptions and to inform monitoring and early warning efforts.

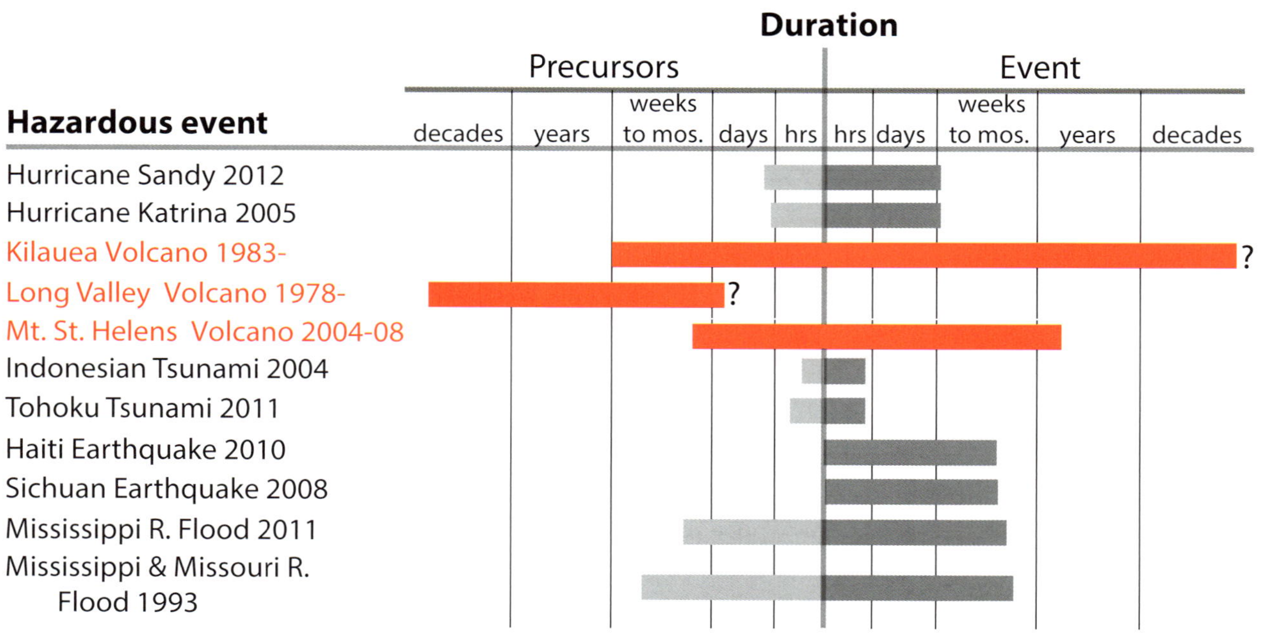

FIGURE 1.3 Duration of precursors and events for selected natural hazards, including hurricanes, volcanic eruptions, earthquakes, and floods.

> **BOX 1.1**
> **Volcano-Related Missions of U.S. Federal Agencies**
>
> In the United States, three federal agencies play a key role in volcano research, monitoring, and/or eruption warning. The U.S. Geological Survey Volcano Hazards Program monitors and studies active and potentially active volcanoes, assesses their hazards, and conducts research on volcano processes to issue forecasts, warnings, and information about volcano hazards to emergency management professionals and the public. The National Science Foundation Division of Earth Sciences supports proposals for research geared toward improving the understanding of the structure, composition, and evolution of the Earth, the life it supports, and the processes that govern the formation and behavior of the Earth's materials. The National Aeronautics and Space Administration's (NASA's) Earth Surface and Interior focus area supports research, analysis, and the use of NASA's unique capabilities and observational resources to better understand core, mantle, and lithospheric structure and dynamics, and interactions between these processes and Earth's fluid envelopes. These studies provide the basic understanding and data products needed to inform the assessment, mitigation, and forecasting of natural hazards, including volcanic eruptions.
>
> SOURCES: https://volcanoes.usgs.gov/vhp/about.html; https://www.nsf.gov/geo/ear/about.jsp; https://science.nasa.gov/earth-science/focus-areas/surface-and-interior.

The roles of the three agencies in advancing volcano science are summarized in Box 1.1.

The committee held four meetings, including an international workshop, to gather information, deliberate, and prepare its report. The report is not intended to be a comprehensive review, but rather to provide a broad overview of the topics listed above. Chapter 2 addresses the opportunities for better understanding the storage, ascent, and eruption of magmas. Chapter 3 summarizes the challenges and prospects for forecasting eruptions and their consequences. Chapter 4 highlights repercussions of volcanic eruptions on a host of other Earth systems. Although not explicitly called out in the four tasks, the interactions between volcanoes and other Earth systems affect the consequences of eruptions, and offer opportunities to improve forecasting and obtain new insights into volcanic processes. Chapter 5 summarizes opportunities to strengthen research in volcano science. Chapter 6 provides overarching conclusions. Supporting material appears in appendixes, including a list of volcano databases (see Appendix A), a list of workshop participants (see Appendix B), biographical sketches of the committee members (see Appendix C), and a list of acronyms and abbreviations (see Appendix D).

Background information on these topics is summarized in the rest of this chapter.

1.2 VOLCANOES IN THE UNITED STATES

The USGS has identified 169 potentially active volcanoes in the United States and its territories (e.g., Marianas), 55 of which pose a high threat or very high threat (Ewert et al., 2005). Of the total, 84 are monitored by at least one seismometer, and only 3 have gas sensors (as of November 2016).[1] Volcanoes are found in the Cascade mountains, Aleutian arc, Hawaii, and the western interior of the continental United States (Figure 1.4). The geographical extent and eruption hazards of these volcanoes are summarized below.

The Cascade volcanoes extend from Lassen Peak in northern California to Mount Meager in British Columbia. The historical record contains only small- to moderate-sized eruptions, but the geologic record reveals much larger eruptions (Carey et al., 1995; Hildreth, 2007). Activity tends to be sporadic (Figure 1.5). For example, nine Cascade eruptions occurred in the 1850s, but none occurred between 1915 and 1980, when Mount St. Helens erupted. Consequently, forecasting eruptions in the Cascades is subject to considerable uncertainty. Over the coming decades, there may be multiple eruptions from several volcanoes or no eruptions at all.

The Aleutian arc extends 2,500 km across the North Pacific and comprises more than 130 active and potentially active volcanoes. Although remote, these volcanoes pose a high risk to overflying aircraft that carry more than 30,000 passengers a day, and are monitored by a combination of ground- and space-based sensors. One or two small to moderate explosive eruptions occur in the Aleutians every year, and very large eruptions occur less frequently. For example, the world's largest eruption of the 20th century occurred approximately 300 miles from Anchorage, in 1912.

In Hawaii, Kilauea has been erupting largely effusively since 1983, but the location and nature of eruptions can vary dramatically, presenting challenges for disaster preparation. The population at risk from large-volume, rapidly moving lava flows on the flanks of the Mauna Loa volcano has grown tremendously in the past few decades (Dietterich and Cashman, 2014), and few island residents are prepared for the even larger magnitude explosive eruptions that are documented in the last 500 years (Swanson et al., 2014).

All western states have potentially active volcanoes, from New Mexico, where lava flows have reached within a few kilometers of the Texas and Oklahoma borders (Fitton et al., 1991), to Montana, which borders the Yellowstone caldera (Christiansen, 1984). These volcanoes range from immense calderas that formed from super-eruptions (Mastin et al., 2014) to small-volume basaltic volcanic fields that erupt lava flows and tephra for a few months to a few decades. Some of these eruptions are monogenic (erupt just once) and pose a special challenge for forecasting. Rates of activity in these distributed volcanic fields are low, with many eruptions during the past few thousand years (e.g., Dunbar, 1999; Fenton, 2012; Laughlin et al., 1994), but none during the past hundred years.

1.3 THE STRUCTURE OF A VOLCANO

Volcanoes often form prominent landforms, with imposing peaks that tower above the surrounding landscape, large depressions (calderas), or volcanic fields with numerous dispersed cinder cones, shield volcanoes, domes, and lava flows. These various landforms reflect the plate tectonic setting, the ways in which those volcanoes erupt, and the number of eruptions. Volcanic landforms change continuously through the interplay between constructive processes such as eruption and intrusion, and modification by tectonics, climate, and erosion. The stratigraphic and structural architecture of volcanoes yields critical information on eruption history and processes that operate within the volcano.

Beneath the volcano lies a magmatic system that in most cases extends through the crust, except during eruption. Depending on the setting, magmas may rise

[1] Personal communication from Charles Mandeville, Program Coordinator, Volcano Hazards Program, U.S. Geological Survey, on November 26, 2016.

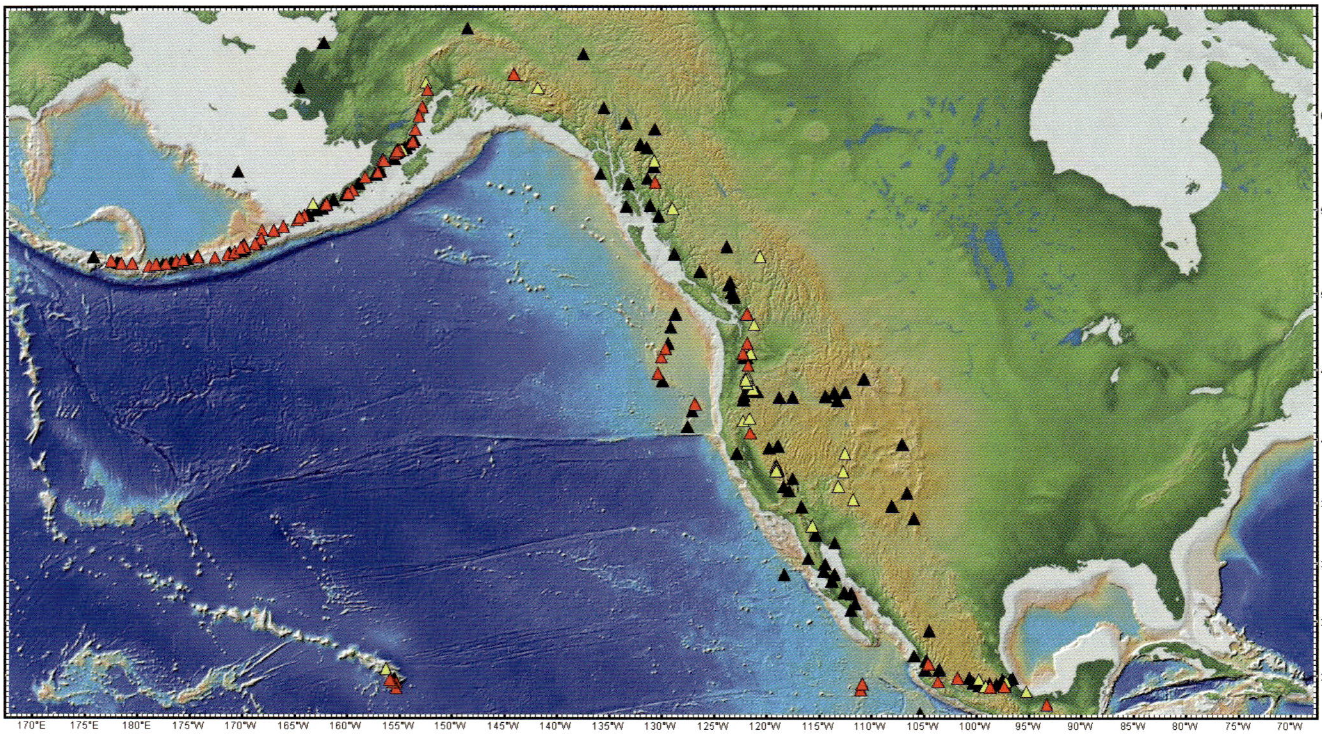

FIGURE 1.4 Map of volcanoes in the United States, Canada, and northern Mexico that have been active in the past 10,000 years, including those that have erupted since 1800 CE (red triangles), in the period of 0 to 1800 AD (yellow triangles), and earlier (black triangles). SOURCES: Data from the Smithsonian Institution's Global Volcanism Program Holocene database (Venzke, 2013), and map created in GeoMapApp (http://www.geomapapp.org).

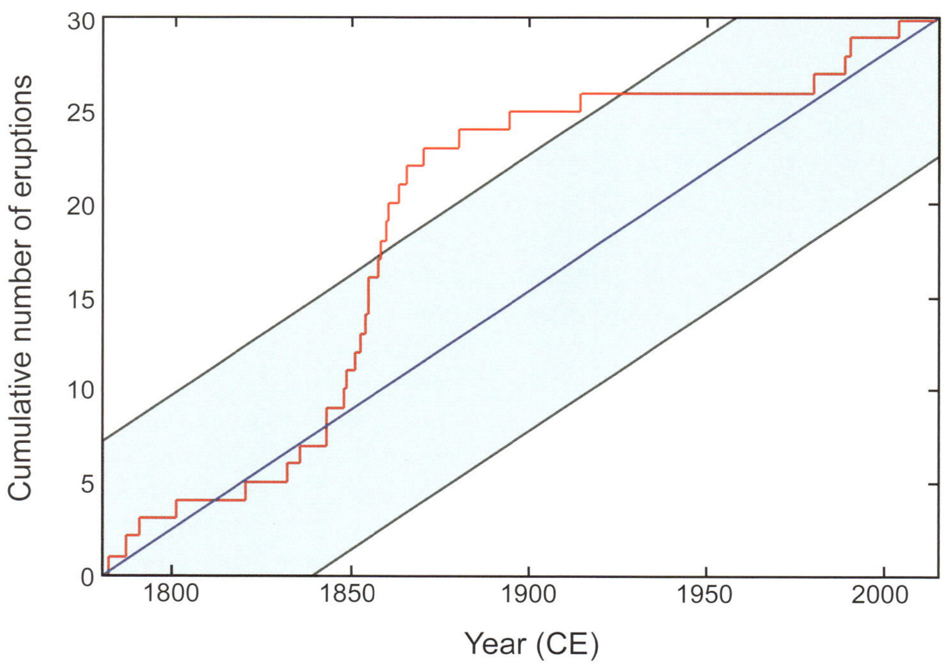

FIGURE 1.5 A total of 30 volcanic eruptions have been documented in the Cascades since the 1786 eruption of Mount Shasta. The cumulative number of eruptions with time (solid line) does not increase at a constant rate. Compared to a model of steady volcanic activity (dashed line), the eruption rate in the Cascades is remarkably variable, with greater than 95 percent confidence (confidence envelope shown by dotted lines). SOURCE: Data from confirmed historical observations reported in the Smithsonian's Global Volcanism Program catalog.

directly from the mantle or be staged in one or more storage regions within the crust before erupting. The uppermost part (within 2–3 km of Earth's surface) often hosts an active hydrothermal system where meteoric groundwater mingles with magmatic volatiles and is heated by deeper magma. Identifying the extent and vigor of hydrothermal activity is important for three reasons: (1) much of the unrest at volcanoes occurs in hydrothermal systems, and understanding the interaction of hydrothermal and magmatic systems is important for forecasting; (2) pressure buildup can cause sudden and potentially deadly phreatic explosions from the hydrothermal system itself (such as on Ontake, Japan, in 2014), which, in turn, can influence the deeper magmatic system; and (3) hydrothermal systems are energy resources and create ore deposits.

Below the hydrothermal system lies a magma reservoir where magma accumulates and evolves prior to eruption. Although traditionally modeled as a fluid-filled cavity, there is growing evidence that magma reservoirs may comprise an interconnected complex of vertical and/or horizontal magma-filled cracks, or a partially molten mush zone, or interleaved lenses of magma and solid material (Cashman and Giordano, 2014). In arc volcanoes, magma chambers are typically located 3–6 km below the surface. The magma chamber is usually connected to the surface via a fluid-filled conduit only during eruptions. In some settings, magma may ascend directly from the mantle without being stored in the crust.

In the broadest sense, long-lived magma reservoirs comprise both eruptible magma (often assumed to contain less than about 50 percent crystals) and an accumulation of crystals that grow along the margins or settle to the bottom of the magma chamber. Physical segregation of dense crystals and metals can cause the floor of the magma chamber to sag, a process balanced by upward migration of more buoyant melt. A long-lived magma chamber can thus become increasingly stratified in composition and density.

The deepest structure beneath volcanoes is less well constrained. Swarms of low-frequency earthquakes at mid- to lower-crustal depths (10–40 km) beneath volcanoes suggest that fluid is periodically transferred into the base of the crust (Power et al., 2004). Tomographic studies reveal that active volcanic systems have deep crustal roots that contain, on average, a small fraction of melt, typically less than 10 percent. The spatial distribution of that melt fraction, particularly how much is concentrated in lenses or in larger magma bodies, is unknown. Erupted samples preserve petrologic and geochemical evidence of deep crystallization, which requires some degree of melt accumulation. Seismic imaging and sparse outcrops suggest that the proportion of unerupted solidified magma relative to the surrounding country rock increases with depth and that the deep roots of volcanoes are much more extensive than their surface expression.

1.4 MONITORING VOLCANOES

Volcano monitoring is critical for hazard forecasts, eruption forecasts, and risk mitigation. However, many volcanoes are not monitored at all, and others are monitored using only a few types of instruments. Some parameters, such as the mass, extent, and trajectory of a volcanic ash cloud, are more effectively measured by satellites. Other parameters, notably low-magnitude earthquakes and volcanic gas emissions that may signal an impending eruption, require ground-based monitoring on or close to the volcanic edifice. This section summarizes existing and emerging technologies for monitoring volcanoes from the ground and from space.

Monitoring Volcanoes on or Near the Ground

Ground-based monitoring provides data on the location and movement of magma. To adequately capture what is happening inside a volcano, it is necessary to obtain a long-term and continuous record, with periods spanning both volcanic quiescence and periods of unrest. High-frequency data sampling and efficient near-real-time relay of information are important, especially when processes within the volcano–magmatic–hydrothermal system are changing rapidly. Many ground-based field campaigns are time intensive and can be hazardous when volcanoes are active. In these situations, telemetry systems permit the safe and continuous collection of data, although the conditions can be harsh and the lifetime of instruments can be limited in these conditions.

Ground-based volcano monitoring falls into four broad categories: seismic, deformation, gas, and thermal monitoring (Table 1.1). Seismic monitoring tools,

TABLE 1.1 Ground-Based Instrumentation for Monitoring Volcanoes

Measurement	Instrument	Purpose
Seismic waves	Geophone	Detect lahars (volcanic mudflows) and pyroclastic density currents
	Short-period seismometer	Locate earthquakes, study earthquake mechanics, and detect unrest
	Broadband seismometer	Study earthquakes, tremor, and long-period earthquakes to quantify rock failure, fluid movement, and eruption progress
	Infrasound detector	Track evolution of near-surface eruptive activity
Geodetic	Classical surveying techniques	Detect deformation over broad areas
	Tiltmeter	Detect subtle pressurization or volumetric sources
	Strainmeter	Detect changing stress distributions
	GNSS/Global Positioning System	Model intrusion locations and sizes, detect ash clouds
	Photogrammetic and structure from motion	Map and identify or measure morphologic changes
	Lidar	Precision mapping, detect ash and aerosol heights
	Radar	Quantify rapid surface movements and velocities of ballistic pyroclasts
Gas	Miniature differential optical absorption spectrometer	Detect sulfur species concentrations and calculate gas flux
	Open-path Fourier transform infrared spectroscopy	Quantify gas concentration ratios
	Ultraviolet imagers	Detect plume sulfur
	Gigenbach-type sampling and multiGAS sensors	Determine chemical and isotopic compositions and make in situ measurements of gas species
	Portable laser spectrometer	Measure stable isotopic ratios of gases
Thermal	Infrared thermal camera	Detect dome growth, lava breakouts, and emissions of volcanic ash and gas
	In situ thermocouple	Monitor fumarole temperatures
Hydrologic	Temperature probe	Detect changes in hydrothermal sources
	Discharge measurements	Detect changes in pressure or permeability
	Sampling for chemical and isotopic composition	Detect magma movement
Potential fields	Gravimeter	Detect internal mass movement
	Self-potential, resistivity	Detect fluids and identify fractures and voids
	Magnetotellurics	3D location of fluids and magma in shallow crust
Other	Cosmic ray muon detector	Tomography
	High-speed camera	Image explosion dynamics
	Drones	Visually observe otherwise inaccessible surface phenomena
	Lightning detection array	Locate lightning and identify ash emissions

including seismometers and infrasound sensors, are used to detect vibrations caused by breakage of rock and movement of fluids and to assess the evolution of eruptive activity. Ambient seismic noise monitoring can image subsurface reservoirs and document changes in wave speed that may reflect stress. changes. Deformation monitoring tools, including tiltmeters, borehole strainmeters, the Global Navigation Satellite System (GNSS, which includes the Global Positioning System [GPS]), lidar, radar, and gravimeters, are used to detect the motion of magma and other fluids in the subsurface. Some of these tools, such as GNSS and lidar, are also used to detect erupted products, including ash clouds, pyroclastic density currents, and volcanic bombs. Gas monitoring tools, including a range of sensors (Table 1.1), and direct sampling of gases and fluids are used to detect magma intrusions and changes in magma–hydrothermal interactions. Thermal monitoring tools, such as infrared cameras, are used to detect dome growth and lava breakouts. Continuous video or photographic observations are also commonly used and, despite their simplicity, most directly document volcanic activity. Less commonly used monitoring technologies, such as self-potential, electromagnetic techniques, and lightning detection are used to constrain fluid movement and to detect ash clouds. In addition, unmanned aerial vehicles (e.g., aircraft and drones) are increasingly being used to collect data. Rapid sample collection and analysis is also becoming more common as a monitoring tool at volcano observatories. A schematic of ground-based monitoring techniques is shown in Figure 1.6.

Monitoring Volcanoes from Space

Satellite-borne sensors and instruments provide synoptic observations during volcanic eruptions when collecting data from the ground is too hazardous or where volcanoes are too remote for regular observation. Repeat-pass data collected over years or decades provide a powerful means for detecting surface changes on active volcanoes. Improvements in instrument sensitivity, data availability, and the computational capacity required to process large volumes of data have led to a dramatic increase in "satellite volcano science."

Although no satellite-borne sensor currently in orbit has been specifically designed for volcano monitoring, a number of sensors measure volcano-relevant

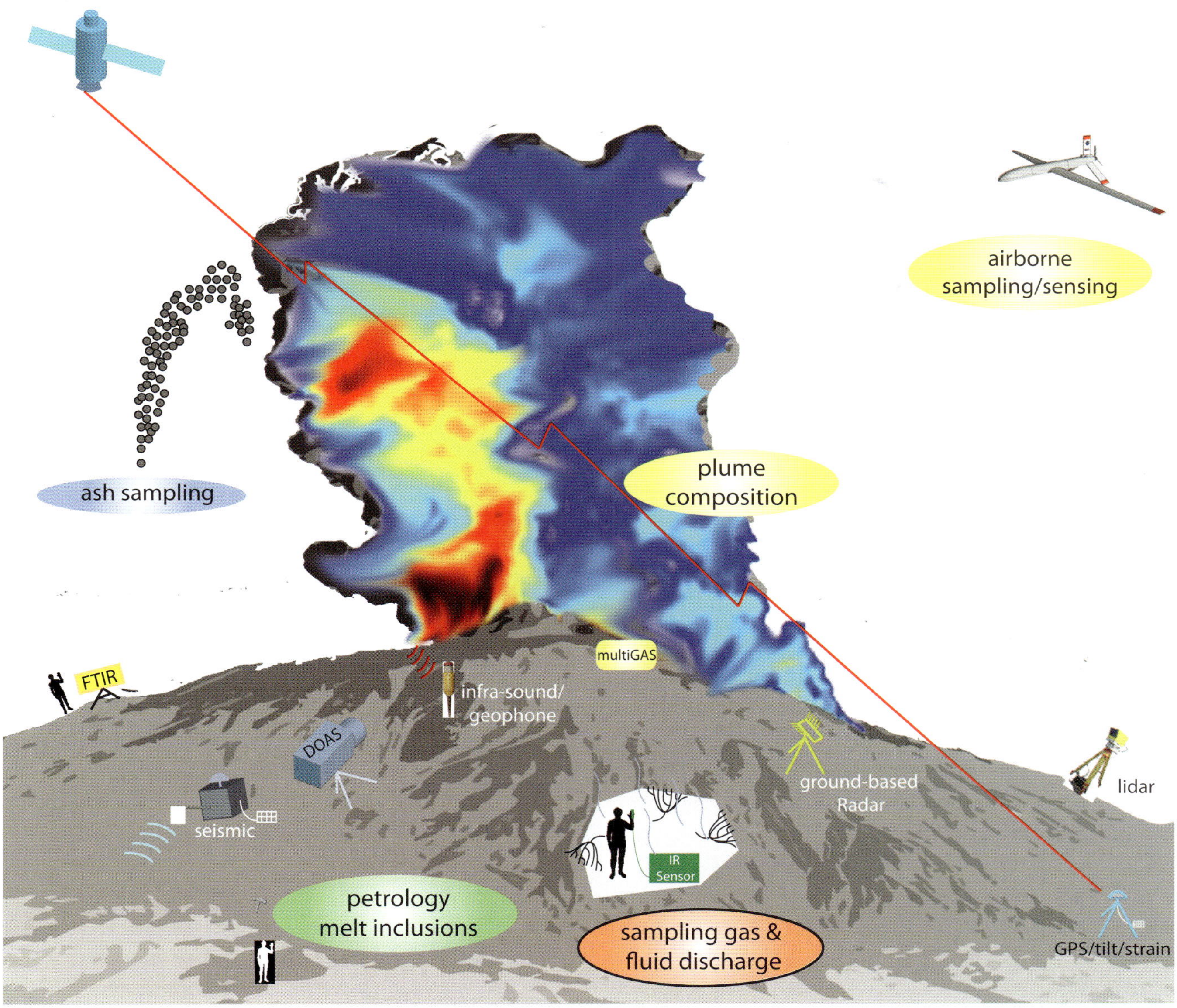

FIGURE 1.6 Some of the tools used to monitor and study volcanoes on or close to the ground. NOTE: DOAS, differential optical absorption spectrometer; FTIR, Fourier Transform Infrared Spectroscopy; GPS, Global Positioning System. Background image is the concentration of SO_2 measured with an ultraviolet camera.

TABLE 1.2 Satellite-Borne Sensor Suite for Volcano Monitoring

Measurement	Purpose	Examples
High-temporal/low-spatial-resolution multispectral thermal infrared	Detect eruptions and map ash clouds	GOES
Low-temporal/moderate-spatial-resolution multispectral thermal infrared	Detect eruptions and map ash clouds with coverage of high latitudes; infer lava effusion rate	AVHRR, MODIS
Low-temporal/high-spatial-resolution multispectral visible infrared	Map detailed surface and plumes; infer lava effusion rate	Landsat, ASTER, Sentinel-2
Hyperspectral ultraviolet	Detect and quantify volcanic SO_2, BrO, and OClO	OMI
Hyperspectral infrared	Detect and quantify volcanic SO_2 and H_2S in nighttime and winter	IASI, AIRS
Microwave limb sounding	Detect volcanic SO_2 and HCl in the upper troposphere and stratosphere	MLS
Visible–near-infrared multiangle imaging	Determine volcanic ash cloud altitudes and plume speed	MISR
Ultraviolet–visible limb scattering	Measure aerosol vertical profiles	OMPS-LP
Ultraviolet–near-infrared solar occultation	Measure stratospheric aerosol	SAGE III
Spaceborne lidar	Develop vertical profiles of volcanic clouds	CALIPSO
Spaceborne W-band radar	Measure volcanic hydrometeors	CloudSat
Multiband (X-, C-, L-band) synthetic aperture radar	Measure deformation globally	Sentinel-1a/b, ALOS-2, COSMO-SkyMed, TerraSAR-X, TanDEM-X, Radarsat-2

NOTE: AIRS, Atmospheric Infrared Sounder; ALOS, Advanced Land Observing Satellite; ASTER, Advanced Spaceborne Thermal Emission and Reflection Radiometer; AVHRR, Advanced Very High Resolution Radiometer; CALIPSO, Cloud-Aerosol Lidar and Infrared Pathfinder Satellite Observation; COSMO-SkyMed, Constellation of Small Satellites for Mediterranean Basin Observation; GOES, Geostationary Operational Environmental Satellite; IASI, Infrared Atmospheric Sounding Interferometer; MISR, Multi-angle Imaging SpectroRadiometer; MLS, Microwave Limb Sounder; MODIS, Moderate Resolution Imaging Spectroradiometer; OMI, Ozone Monitoring Instrument; OMPS, Ozone Mapping and Profiler Suite; SAGE, Stratospheric Aerosol and Gas Experiment.

parameters, including heat flux, gas and ash emissions, and deformation (Table 1.2). Thermal infrared data are used to detect eruption onset and cessation, calculate lava effusion rates, map lava flows, and estimate ash column heights during explosive eruptions. In some cases, satellites may capture thermal precursors to eruptions, although low-temperature phenomena are challenging to detect. Both high-temporal/low-spatial-resolution (geostationary orbit) and high-spatial/low-temporal-resolution (polar orbit) thermal infrared observations are needed for global volcano monitoring.

Satellite-borne sensors are particularly effective for observing the emission and dispersion of volcanic gas and ash plumes in the atmosphere. Although several volcanic gas species can be detected from space (including SO_2, BrO, OClO, H_2S, HCl, and CO; Carn et al., 2016), SO_2 is the most readily measured, and it is also responsible for much of the impact of eruptions on climate. Satellite measurements of SO_2 are valuable for detecting eruptions, estimating global volcanic fluxes and recycling of other volatile species, and tracking volcanic clouds that may be hazardous to aviation in near real time. Volcanic ash cloud altitude is most accurately determined by spaceborne lidar, although spatial coverage is limited. Techniques for measuring volcanic CO_2 from space are under development and could lead to earlier detection of preeruptive volcanic degassing.

Interferometric synthetic aperture radar (InSAR) enables global-scale background monitoring of volcano deformation (Figure 1.7). InSAR provides much higher spatial resolution than GPS, but lower accuracy and temporal resolution. However, orbit repeat times will diminish as more InSAR missions are launched, such as the European Space Agency's recently deployed Sentinel-1 satellite and the NASA–Indian Space Research Organisation synthetic aperture radar mission planned for launch in 2020.

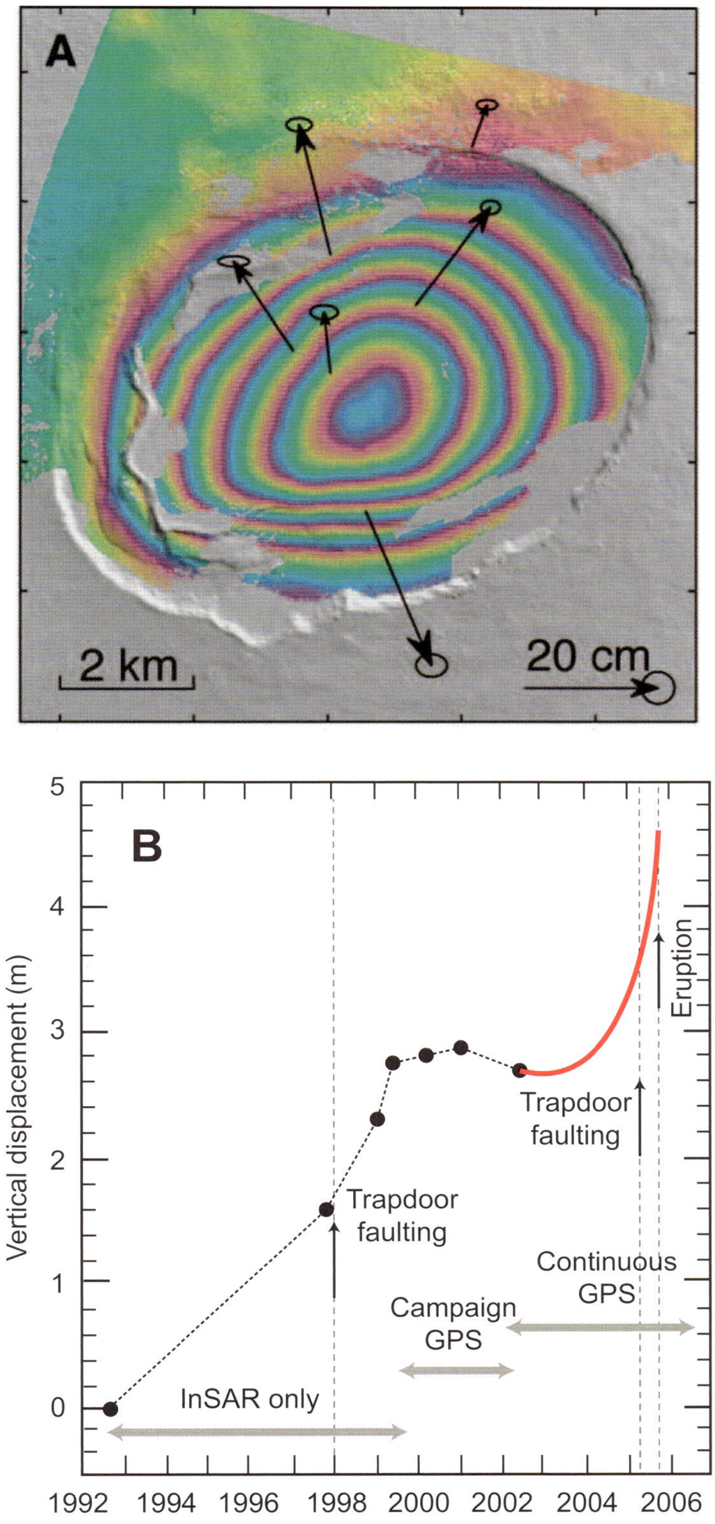

FIGURE 1.7 Interferometric synthetic aperture radar (InSAR) measurements of the Sierra Negra volcano, Galapagos. (A) Uplift from February 12, 2004, to January 27, 2005. Each fringe (i.e., the repetition of a color) represents a 10-cm range change. (B) Uplift history of center of caldera at Sierra Negra from 1992 to 2006, determined from InSAR and GPS. The volcano inflated nearly 5 m before it erupted on October 22, 2005. SOURCE: Modified from Chadwick et al. (2006).

1.5 ERUPTION BEHAVIOR

Eruptions range from violently explosive to gently effusive, from short lived (hours to days) to persistent over decades or centuries, from sustained to intermittent, and from steady to unsteady (Siebert et al., 2015). Eruptions may initiate from processes within the magmatic system (Section 1.3) or be triggered by processes and properties external to the volcano, such as precipitation, landslides, and earthquakes. The eruption behavior of a volcano may change over time. No classification scheme captures this full diversity of behaviors (see Bonadonna et al., 2016), but some common schemes to describe the style, magnitude, and intensity of eruptions are summarized below.

Eruption Magnitude and Intensity

The size of eruptions is usually described in terms of total erupted mass (or volume), often referred to as magnitude, and mass eruption rate, often referred to as intensity. Pyle (2015) quantified magnitude and eruption intensity as follows:

$$\text{magnitude} = \log_{10} (\text{mass, in kg}) - 7, \text{ and}$$
$$\text{intensity} = \log_{10} (\text{mass eruption rate, in kg/s}) + 3.$$

The Volcano Explosivity Index (VEI) introduced by Newhall and Self (1982) assigns eruptions to a VEI class based primarily on measures of either magnitude (erupted mass or volume) or intensity (mass eruption rate and/or eruption plume height), with more weight given to magnitude. The VEI classes are summarized in Figure 1.8. The VEI classification is still in use, despite its many limitations, such as its reliance on only a few types of measurements and its poor fit for small to moderate eruptions (see Bonadonna et al., 2016).

Smaller VEI events are relatively common, whereas larger VEI events are exponentially less frequent (Siebert et al., 2015). For example, on average about three VEI 3 eruptions occur each year, whereas there is a 5 percent chance of a VEI 5 eruption and a 0.2 percent chance of a VEI 7 (e.g., Crater Lake, Oregon) event in any year.

Eruption Style

The style of an eruption encompasses factors such as eruption duration and steadiness, magnitude, gas flux, fountain or column height, and involvement of magma and/or external source of water (phreatic and phreatomagmatic eruptions). Eruptions are first divided into effusive (lava producing) and explosive (pyroclast producing) styles, although individual eruptions can be simultaneously effusive and weakly explosive, and can pass rapidly and repeatedly between eruption styles. Explosive eruptions are further subdivided into styles that are sustained on time scales of hours to days and styles that are short lived (Table 1.3).

Classification of eruption style is often qualitative and based on historical accounts of characteristic eruptions from type-volcanoes. However, many type-volcanoes exhibit a range of eruption styles over time (e.g., progressing between Strombolian, Vulcanian, and Plinian behavior; see Fee et al., 2010), which has given rise to terms such as subplinian or violent Strombolian.

1.6 ERUPTION HAZARDS

Eruption hazards are diverse (Figure 1.9) and may extend more than thousands of kilometers from an active volcano. From the perspective of risk and impact, it is useful to distinguish between near-source and distal hazards. Near-source hazards are far more unpredictable than distal hazards.

Near-source hazards include those that are airborne, such as tephra fallout, volcanic gases, and volcanic projectiles, and those that are transported laterally on or near the ground surface, such as pyroclastic density currents, lava flows, and lahars. Pyroclastic density currents are hot volcanic flows containing mixtures of gas and micron- to meter-sized volcanic particles. They can travel at velocities exceeding 100 km per hour. The heat combined with the high density of material within these flows obliterates objects in their path, making them the most destructive of volcanic hazards. Lava flows also destroy everything in their path, but usually move slowly enough to allow people to get out of the way. Lahars are mixtures of volcanic debris, sediment, and water that can travel many tens of kilometers along valleys and river channels. They may be triggered during an eruption by interaction between volcanic prod-

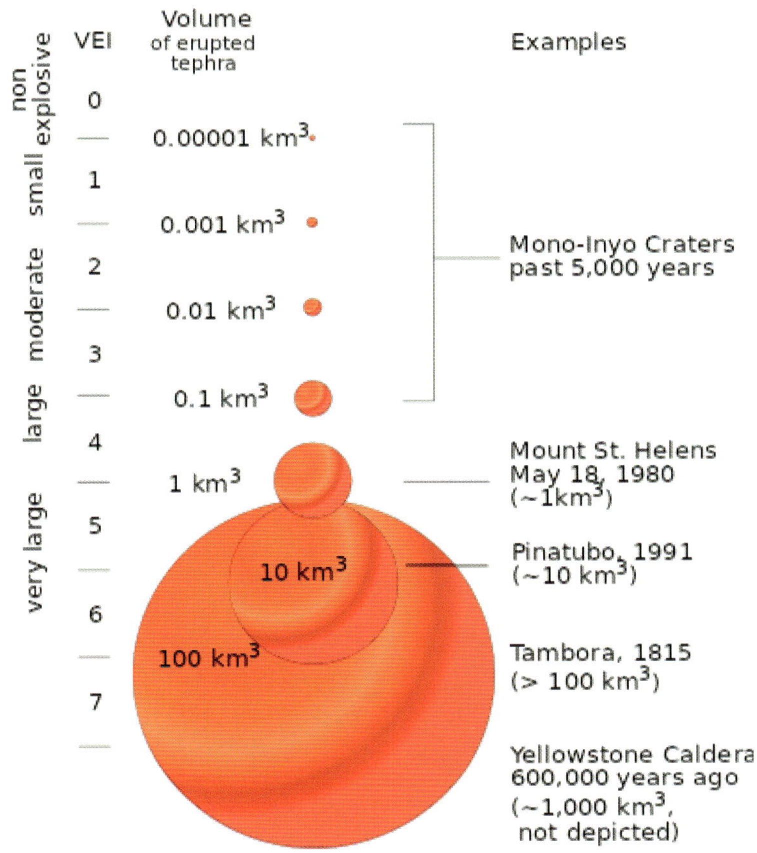

FIGURE 1.8 Definition of the Volcanic Explosivity Index (VEI) scale (Newhall and Self, 1982). While this classification of magnitude does not capture the diversity of eruption features it is a starting point for characterization and comparison of the volumes of magma erupted in explosive eruptions. SOURCE: USGS.

TABLE 1.3 Characteristics of Different Eruption Styles

Eruption Style	Characteristics
Hawaiian	Sustained fountaining of magmatic gas and pyroclasts (up to ~1,000 m) often generating clastogenic, gas-charged lava flows from single vents or from fissures
Strombolian	Short-duration, low-vigor, episodic, small (<100s of meters) explosions driven by escape of pockets of gas and ejecting some bombs and spatter
Vulcanian	Short-duration, moderately vigorous, magma-fragmenting explosions producing ash-rich columns that may reach heights >1,000 m
Surtseyan	Short duration, weak phreatomagmatic explosive eruptions where fluid magma interacts with standing water
Phreatoplinian	Prolonged powerful phreatomagmatic explosions where viscous magma interacts with surface water or groundwater
Dome collapse	Dome collapse pyroclastic flows occur at unstable gas-charged domes either with an explosive central column eruption (e.g., Mount Pelee) or without (e.g., Unzen, Montserrat, and Santiaguito)
Plinian	Very powerful, sustained eruptions with columns reaching the stratosphere (>15 km) and sometimes generating large pyroclastic density currents from collapsing eruption columns

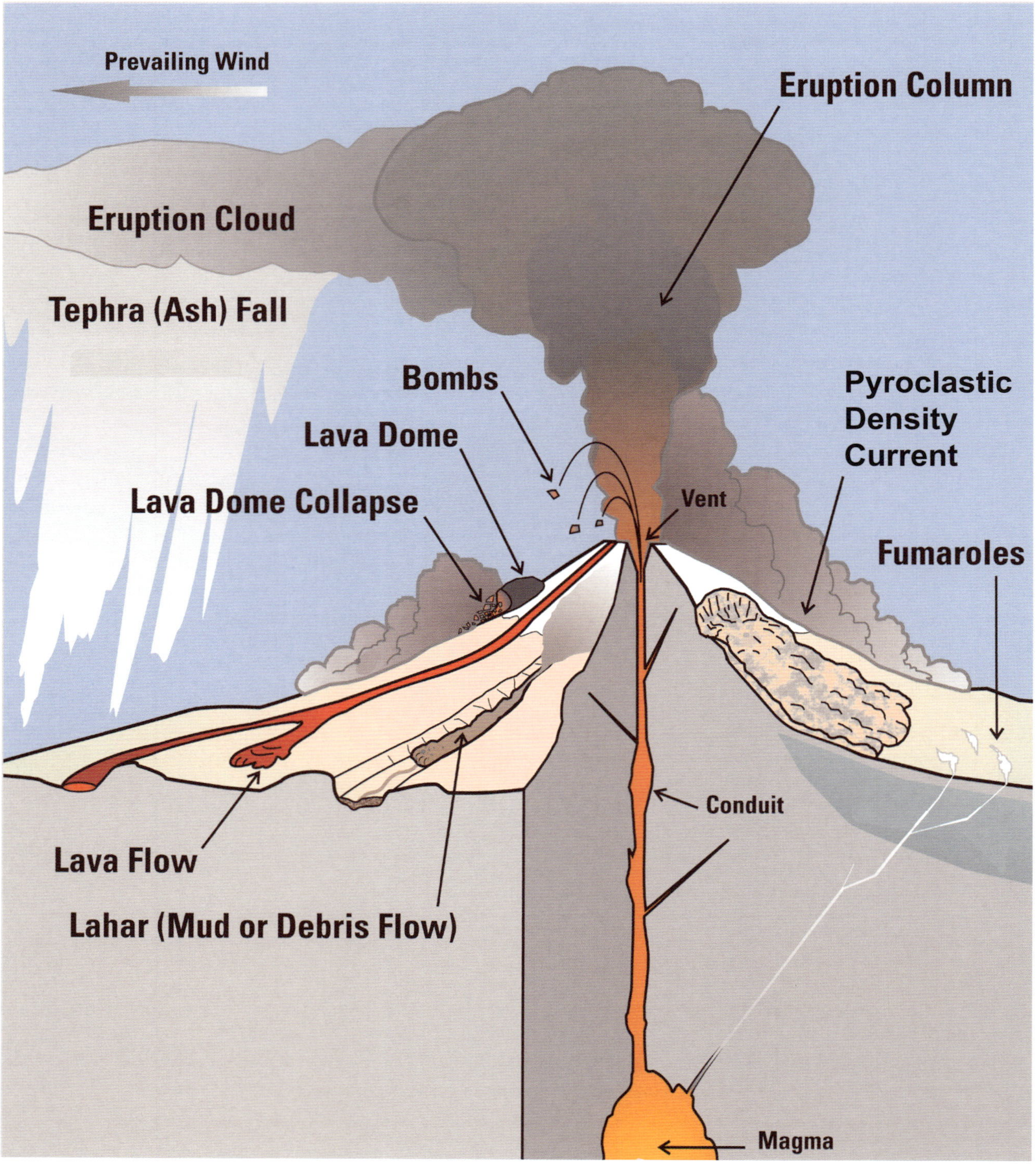

FIGURE 1.9 Volcanoes have impacts near and far from the volcano, before, during, and long after eruption. Near-source hazards are acute events that operate on very short time frames, on and close to the volcano, with limited warning time. Distal hazards include flooding and sedimentation over extended areas, airborne ash, and fallout of tephra downwind of the volcano. SOURCE: USGS.

ucts and snow, ice, rain, or groundwater. Lahars can be more devastating than the eruption itself. Ballistic blocks are large projectiles that typically fall within 1–5 km from vents.

The largest eruptions create distal hazards. Explosive eruptions produce plumes that are capable of dispersing ash hundreds to thousands of kilometers from the volcano. The thickness of ash deposited depends on the intensity and duration of the eruption and the wind direction. Airborne ash and ash fall are the most severe distal hazards and are likely to affect many more people than near-source hazards. They cause respiratory problems and roof collapse, and also affect transport networks and infrastructure needed to support emergency response. Volcanic ash is a serious risk to air traffic. Several jets fully loaded with passengers have temporarily lost power on all engines after encountering dilute ash clouds (e.g., Guffanti et al., 2010). Large lava flows, such as the 1783 Laki eruption in Iceland, emit volcanic gases that create respiratory problems and acidic rain more than 1,000 km from the eruption. Observed impacts of basaltic eruptions in Hawaii and Iceland include regional volcanic haze ("vog") and acid rain that affect both agriculture and human health (e.g., Thordarson and Self, 2003) and fluorine can contaminate grazing land and water supplies (e.g., Cronin et al., 2003). Diffuse degassing of CO_2 can lead to deadly concentrations with fatal consequences such as occurred at Mammoth Lakes, California, or cause lakes to erupt, leading to massive CO_2 releases that suffocate people (e.g., Lake Nyos, Cameroon).

Secondary hazards can be more devastating than the initial eruption. Examples include lahars initiated by storms, earthquakes, landslides, and tsunamis from eruptions or flank collapse; volcanic ash remobilized by wind to affect human health and aviation for extended periods of time; and flooding because rain can no longer infiltrate the ground.

1.7 MODELING VOLCANIC ERUPTIONS

Volcanic processes are governed by the laws of mass, momentum, and energy conservation. It is possible to develop models for magmatic and volcanic phenomena based on these laws, given sufficient information on mechanical and thermodynamic properties of the different components and how they interact with each other. Models are being developed for all processes in volcanic systems, including melt transport in the mantle, the evolution of magma bodies within the crust, the ascent of magmas to the surface, and the fate of magma that erupts effusively or explosively.

A central challenge for developing models is that volcanic eruptions are complex multiphase and multicomponent systems that involve interacting processes over a wide range of length and time scales. For example, during storage and ascent, the composition, temperature, and physical properties of magma and host rocks evolve. Bubbles and crystals nucleate and grow in this magma and, in turn, greatly influence the properties of the magmas and lavas. In explosive eruptions, magma fragmentation creates a hot mixture of gas and particles with a wide range of sizes and densities. Magma also interacts with its surroundings: the deformable rocks that surround the magma chamber and conduit, the potentially volatile groundwater and surface water, a changing landscape over which pyroclastic density currents and lava flows travel, and the atmosphere through which eruption columns rise.

Models for volcanic phenomena that involve a small number of processes and that are relatively amenable to direct observation, such as volcanic plumes, are relatively straightforward to develop and test. In contrast, phenomena that occur underground are more difficult to model because there are more interacting processes. In those cases, direct validation is much more challenging and in many cases impossible. Forecasting ash dispersal using plume models is more straightforward and testable than forecasting the onset, duration, and style of eruption using models that seek to explain geophysical and geochemical precursors. In all cases, however, the use of even imperfect models helps improve the understanding of volcanic systems.

Modeling approaches can be divided into three categories:

1. Reduced models make simplifying assumptions about dynamics, heat transfer, and geometry to develop first-order explanations for key properties and processes, such as the velocity of lava flows and pyroclastic density currents, the height of eruption columns, the magma chamber size and depth, the dispersal of tephra, and the ascent of magma in conduits. Well-calibrated or tested reduced models offer a straightforward ap-

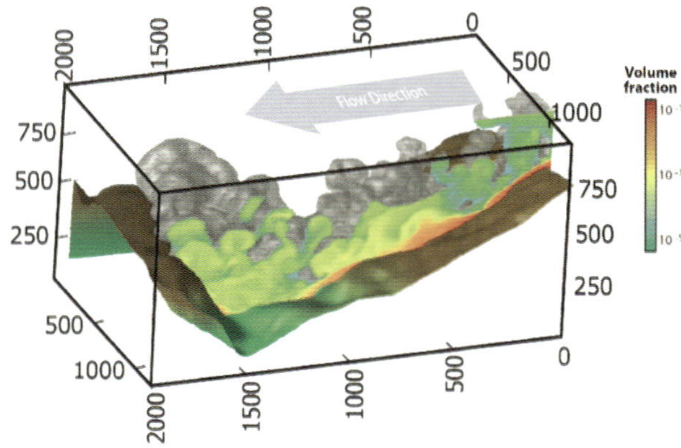

FIGURE 1.10 Multiphase simulation of a pyroclastic density current from the 2006 eruption of Tungurahua, Ecuador, showing the interaction of the current with topography and the formation of dilute, turbulent eddies. The outer grey surface depicts a very dilute condition (10^{-5} volume fraction of particles) similar to what one would observe visually as the outer edge of the current. The cross section enables one to "see" inside the flow. The cross-section colors indicate volume fraction with a lower bed load of concentrated, large clasts (red, volume fraction >0.1) seen in the channelized, upstream portion of the flow. The axes are in meters. Models such as these can be used to understand how flows are influenced by topography and modify landscapes. SOURCE: Modified from Dufek (2016).

proach for combining observations and models in real time in an operational setting (e.g., ash dispersal forecasting for aviation safety). Models may not need to be complex if they capture the most important processes, although simplifications require testing against more comprehensive models and observations.

2. Multiphase and multiphysics models improve scientific understanding of complex processes by invoking fewer assumptions and idealizations than reduced models (Figure 1.10), but at the expense of increased complexity and computational demands. They also require additional components, such as a model for how magma in magma chambers and conduits deforms when stressed; a model for turbulence in pyroclastic density currents and plumes; terms that describe the thermal and mechanical exchange among gases, crystals, and particles; and a description of ash aggregation in eruption columns. A central challenge for multiphysics models is integrating small-scale processes with large-scale dynamics. Many of the models used in volcano science build on understanding developed in other science and engineering fields and for other ap-

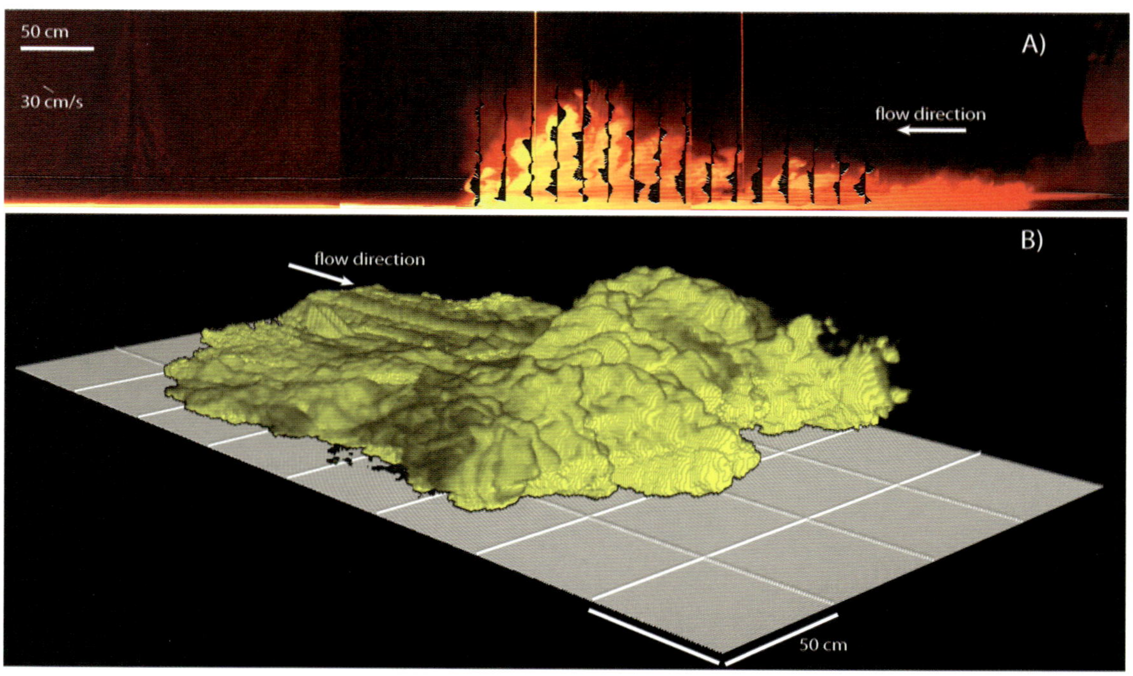

FIGURE 1.11 Dilute density current experiments comprising 20-micron talc particles suspended in air. (A) Vertical slice through a current oriented parallel to the flow direction, 20 seconds after initiation. Color indicates particle concentration (yellow is highest). The two-dimensional plane illuminated by a laser sheet allows velocity and particle concentration to be measured. (B) Three-dimensional reconstruction of a current obtained using a swept laser sheet and high-speed camera. Together with temperature data, these observations allow air entrainment into the currents to be measured. Entrainment controls the distance flows travel. SOURCE: Courtesy of Ben Andrews, Smithsonian Institution.

plications. Multiphysics and multiscale models benefit from rapidly expanding computational capabilities.

3. Laboratory experiments simulate processes for which the geometry and physical and thermal processes and properties can be scaled (Mader et al., 2004). Such experiments provide insights on fundamental processes, such as crystal dynamics in flowing magmas, entrainment in eruption columns, propagation of dikes, and sedimentation from pyroclastic density currents (Figure 1.11). Experiments have also been used successfully to develop the subsystem models used in numerical simulations, and to validate computer simulations for known inputs and properties.

The great diversity of existing models reflects to a large extent the many interacting processes that operate in volcanic eruptions and the corresponding simplifying assumptions currently required to construct such models. The challenge in developing models is often highlighted in discrepancies between models and observations of natural systems. Nevertheless, eruption models reveal essential processes governing volcanic eruptions, and they provide a basis for interpreting measurements from prehistoric and active eruptions and for closing observational gaps. Mathematical models offer a guide for what observations will be most useful. They may also be used to make quantitative and testable predictions, supporting forecasting and hazard assessment.

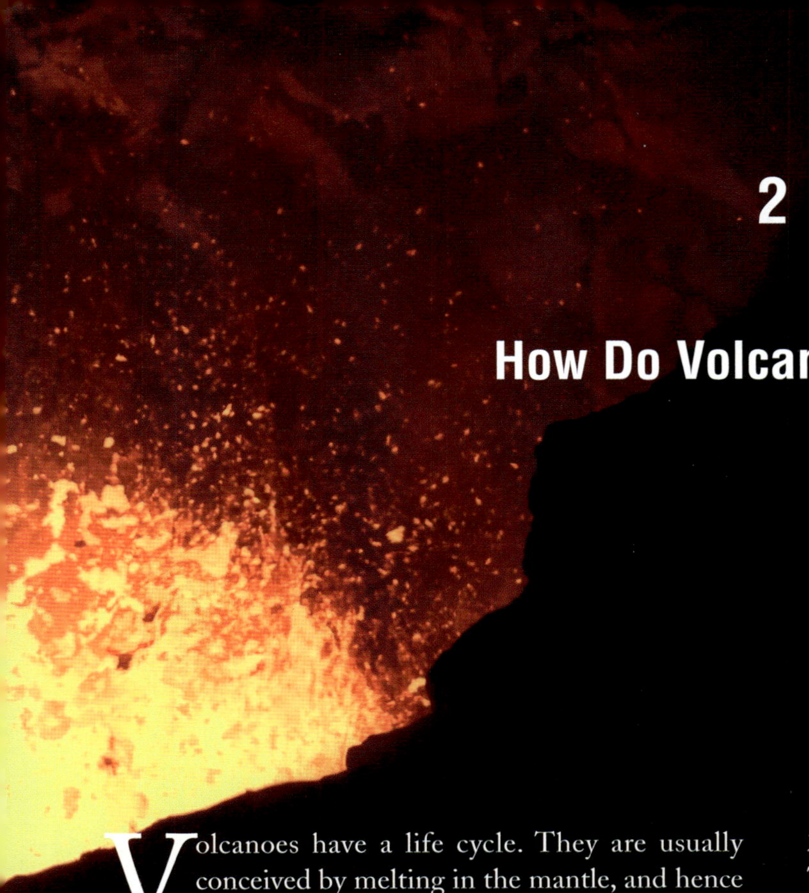

2

How Do Volcanoes Work?

Volcanoes have a life cycle. They are usually conceived by melting in the mantle, and hence their locations are controlled by plate tectonics and mantle convection (Box 2.1). The silicate melts can then ascend to the surface directly, or accumulate in the crust where their volumes and compositions change as they interact with their surroundings. Magma can have a complex history underground. The eruption of magmas creates volcanoes and affects other surface environments such as the hydrosphere and atmosphere. The interactions between melting, storage, accumulation, eruption, and geologic setting give rise to the great diversity seen in eruptions and volcanic landforms.

Each volcano has its own distinct life cycle, often with multiple episodes of repose, unrest, and eruption. Yet the evolution and eruption of all volcanoes are still governed by the same set of processes intrinsic to the magma and influenced by geologic setting. Thus a central challenge to understanding how magma is generated, is stored, ascends, and erupts is to disentangle the unique features of the birth, life, and death of each volcano from the common processes governing their life cycles. This chapter summarizes current understanding of how volcanoes work and identifies key questions and research priorities in three areas: (1) processes that move and store magma beneath volcanoes; (2) how eruptions begin, evolve, and end; and (3) how a volcano erupts.

2.1 HOW ARE MAGMAS STORED AND TRANSPORTED IN THE CRUST?

The path magma takes to the surface is poorly understood. Magma is buoyant and rises through the crust, sometimes erupting at the surface. At hotspots such as Iceland, Hawaii, and some volcanoes in the western United States, magma can ascend directly from the mantle to the surface. But much of the time, magma stalls and forms reservoirs that later erupt or freeze (Figure 2.1). Magma cools because the crust is cooler than the magma, and magma decompresses as it rises. Cooling leads to crystallization and increased viscosity. Decompression may lead to increased buoyancy due to the formation of bubbles from gas originally dissolved in the melt. A loss of volatiles also increases the melt viscosity. Storage and ascent are influenced by the mechanical properties and behavior of the crust, including its ability to deform, flow, or fracture. These properties evolve over the life cycle of the volcano. The competing drivers that force magma to rise and also to resist movement are partly what makes magma movement and eruption so difficult to forecast (Melnik and Sparks, 2006). Will magma stall because of increased viscosity? Or will bubble expansion accelerate magma to the surface in an explosive eruption? The processes that move and store magma are thus fundamental not only to the transfer of mass from the interior to the

BOX 2.1
How and Where Are Magmas Born?

Magmas are generated by melting rocks. Melting can be induced by three mechanisms: heating, lowering pressure, or adding a contaminant (e.g., water) to reduce the melting temperature. Heat loss from Earth's interior drives mantle convection at speeds of centimeters per year, creating tectonic plates at the surface and recycling those plates into the deep Earth. Melting of rocks to produce magmas is governed by the large-scale dynamics produced by those plates and mantle convection, leading to a global correlation of the location of volcanoes and plate boundaries (see figure). Magmas erupted at the surface thus provide a window into the dynamics and evolution of Earth's interior.

At midocean ridges, mantle that rises under spreading plates melts by decompression to produce basaltic magma and the oceanic crust. Some volcanism can also occur where continental crust rifts and mantle upwells and melts, such as in the western United States and East Africa. The convecting mantle can also produce basaltic magma by decompression and create rising plumes of hot mantle away from plate boundaries. These plumes, which may tap hotter-than-normal mantle, are responsible for many of the ocean-island volcanoes, such as the Hawaiian islands.

The other main mechanism for producing magma is related to subduction of tectonic plates. During this process, fluids released from the downgoing plate serve as the contaminating agents that lower the melting temperature of mantle rocks, generating water-rich magmas. Although decompression melting at divergent plate boundaries produces most of the magmas that erupt on Earth today (greater than 75 percent), the majority of the volcanoes located on continents result from melts produced through subduction. For example, most of the volcanoes along the western margin of North America and South America, as well as the volcanoes of the Aleutian Islands in Alaska and around the Pacific (the "Ring of Fire") are produced through this process. Subduction zones generate most of the explosive volcanism that has occurred in the past 100,000 years.

Once magmas are produced in the mantle, their buoyancy relative to their surroundings drives their ascent toward the surface. Ultimately, the fate of the magmas (to erupt or stall in the crust) and the rate of ascent are heavily influenced by the volatile budget of the magmas, their temperature, and their supply rate, which are originally set during the melting process. For example, the addition of volatiles during melting at subduction zones leads to magmas that have different physical properties than their drier cousins, ultimately influencing their eruptive style and vigor.

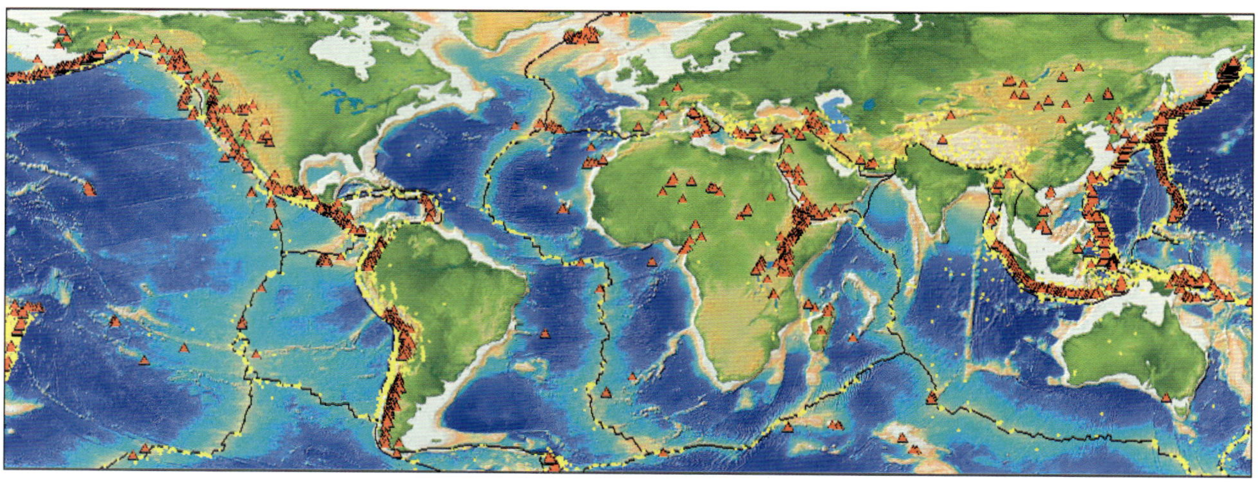

FIGURE Map showing the distribution of volcanoes that have erupted in the last 10,000 years (red triangles; Global Volcanism Program), earthquakes with magnitude greater than 5 since 1990 (yellow dots; http://earthquake.usgs.gov/data/comcat), and plate boundaries (green lines).

exterior of Earth over its history, but also to the style, intensity, magnitude, and duration of volcanic eruptions (Acocella, 2014).

Most volcanoes are not continuously active but spend much of their lifetime at rest, sometimes for thousands of years before erupting again. Prior to an eruption, the movement of magma and fluids may cause earthquakes beneath the volcano, gas emission into the atmosphere or aquifers, and uplift of the ground surface. Importantly, however, these signs of volcanic unrest do not always presage an eruption. Even at rest, volcanoes are unstable landforms, prone to rapid ero-

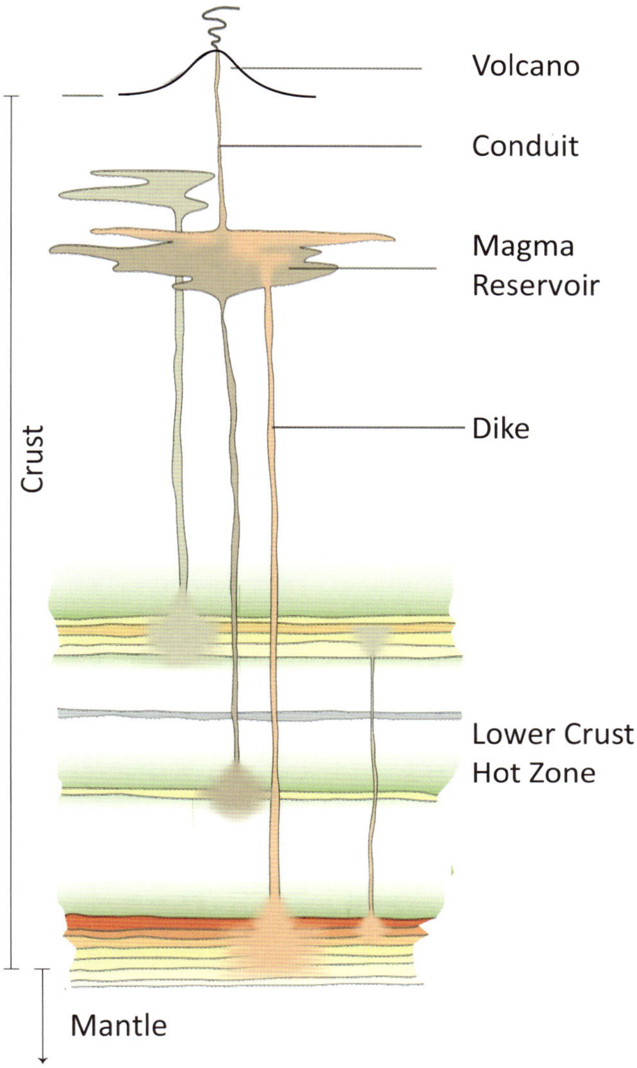

FIGURE 2.1 The volcano is the surface part of a network of magma storage and transport through the entire crust, with thickness between 10 and 60 km, where magma interacts with its surroundings. The reservoirs in which magma accumulates and evolves often have complex geometries and interact with each other. Transport pathways may extend directly from the mantle to the surface, or magma may be stored in (and possibly never erupt from) one or more reservoirs in the crust. Vertical ascent pathways to the surface may be open only during eruption. SOURCE: Annen et al. (2006). Reproduced with permission from C. Annen et al. The Genesis of Intermediate and Silicic Magmas in Deep Crustal Hot Zones. Journal of Petrology (2006) 47 (3): 505-539. Published by Oxford University Press online at: https://academic.oup.com/petrology/article/47/3/505/1536924/The-Genesis-of-Intermediate-and-Silicic-Magmas-in?searchresult=1. Not covered by any Creative Commons or Open-Access License allowing onward reuse. For permissions please contact journals.permissions@oup.com.

sion and collapse, creating hazards even in the absence of eruption. Thus, the life cycle of volcanoes involves alternating periods of repose and unrest punctuated by eruption. We still do not know which signs of unrest signal magma versus gas movement. Which are precursors to eruption? What is normal background activity of volcanoes over their life cycle?

Detecting Magma Under the Ground

A major challenge in studying magma movements is that we are unable to see directly where magma is stored and how it moves under the volcano. From this perspective, it is fortunate that magma does not move quietly or gently. Instead, magma, its movement, and the stresses it generates in the surrounding rock can be detected using the deformation of Earth's surface, the location of earthquakes (Figure 2.2), the frequency of ground shaking, the direction of displacement on faults, and the way seismic waves propagate through the crust. For example, as magma rises and pressurizes subvolcanic reservoirs, it causes the ground surface to rise (inflate). Ground deformation at the scale of millimeters can be sensed with satellite radar, Global Positioning System (GPS), tiltmeters, and strainmeters. Using these tools, it is possible to constrain a combination of the depth and shape of magma reservoirs. Some erupting volcanoes are observed to "breathe," as the subsurface inflates prior to eruption, then deflates after the eruption due to magma removal (Figure 2.3). However, some breathing cycles are not coupled to eruptions and may originate in the hydrothermal systems surrounding the magma.

Geophysical imaging provides additional observational constraints on the location, geometry, and state of magma stored in the crust. Recent advances in seismic tomography, including combinations of body and surface waves (e.g., Syracuse et al., 2015), full waveform inversion and imaging of reflected signals (Arnulf et al., 2014), and the use of the ambient seismic wave field, allow four-dimensional (space and time) imaging of crustal magma reservoirs (e.g., Brenguier et al., 2016; Greenfield et al., 2016; Huang et al., 2015; Jaxybulatov et al., 2014). Tomographic images reveal seismic wave speeds which, when combined with experimental data, yield estimates of temperature and the percentage of partial melt present (e.g., Figure 2.4). Attenuation tomography measures the decay in seismic

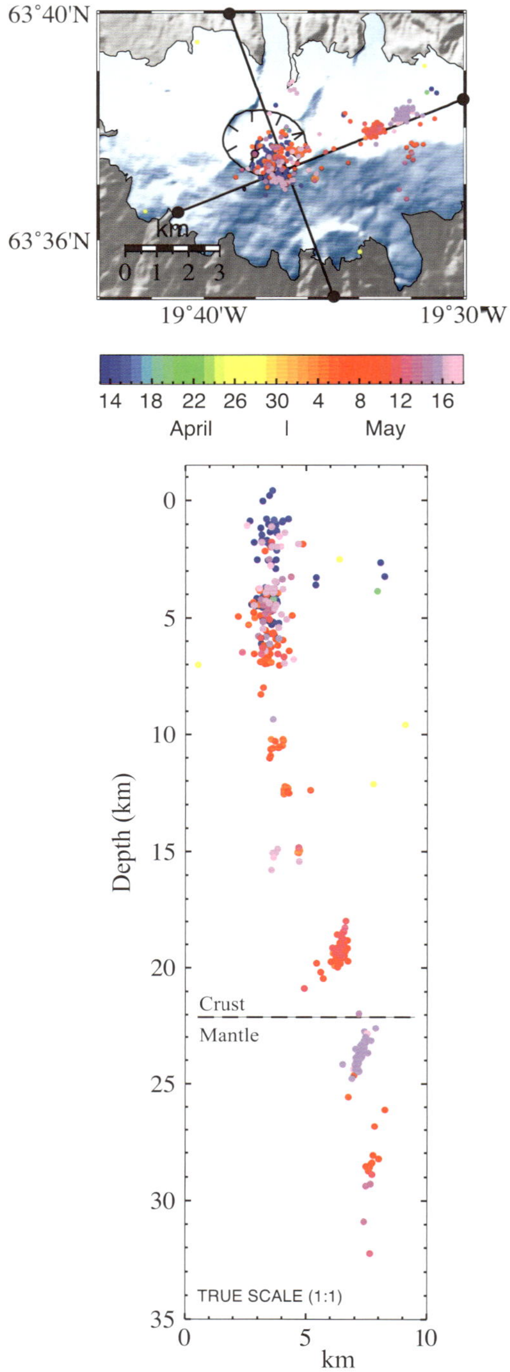

FIGURE 2.2 Earthquake locations are used to detect magma in the subsurface and track the ascent of magma. During the 2010 eruption of Eyjafjallajökull volcano, Iceland, seismicity (color-coded by date) extended into the mantle, revealing an interconnected magmatic system beneath the erupting volcano. This pattern of data suggests rapid ascent of magma from mantle depths during the course of a single eruption. SOURCE: Modified from Tarasewicz et al. (2012).

wave amplitude and can be particularly sensitive to the presence of melt (e.g., De Siena et al., 2014; Lin et al., 2015). Magnetotelluric surveys provide complementary information on the presence of melt and hydrothermal reservoirs (e.g., Desissa et al., 2013; Peacock et al., 2016). Joint inversion of magnetotelluric, seismic, gravity, and other geophysical data has the potential to tighten bounds on subsurface magma systems because the different data types are sensitive to distinct physical properties of the magma and host rocks. Including petrological constraints on composition, temperature, and volatile content (e.g., Comeau et al., 2016) reduces the uncertainty and makes interpretations more physically plausible.

Gases emitted before and during eruption, diffusively and from vents, provide clues about the locations, masses, and histories of magma in the crust (e.g., Aiuppa et al., 2011; Lowenstern et al., 2014). Real-time sampling of the erupted material provides information about syn-eruptive changes in intrinsic magma properties, such as temperature, viscosity, density, and pre-eruptive gas content (e.g., Burgisser et al., 2012). Importantly, these different techniques provide different and complementary images of the subsurface (e.g., Chiodini et al., 2012).

Geologic and geochemical tools have also been developed to study magma underground. Drilling has provided rare samples of magma near the surface, as well as hot rocks and fluids that indicate the temperatures and permeability of the shallowest subvolcanic regions (Elders et al., 2011; Friðleifsson et al., 2013; Marsh et al., 2008; Mortensen et al., 2014; Nakada et al., 2005; Zierenberg et al., 2012). The occurrence of volcanic deposits that are very large and chemically monotonous attests to the existence of large, homogeneous magma reservoirs that supply some giant eruptions (e.g., Bachmann and Bergantz, 2003). The duration of storage, the rate of movement, pressure, and temperature are potentially recorded in erupted crystals (Kahl et al., 2013; Putirka and Tepley, 2008), pockets of melt (now glass) trapped in crystals (Kent, 2008; Sides et al., 2014; Wallace, 2005), and the sizes and shapes of bubbles and crystals (Hammer, 2008; Marsh, 2007; Sable et al., 2006). Interpreting these records is not always straightforward. In particular, the depth of magma storage is difficult to determine and calls for experimental calibration of new crystal and

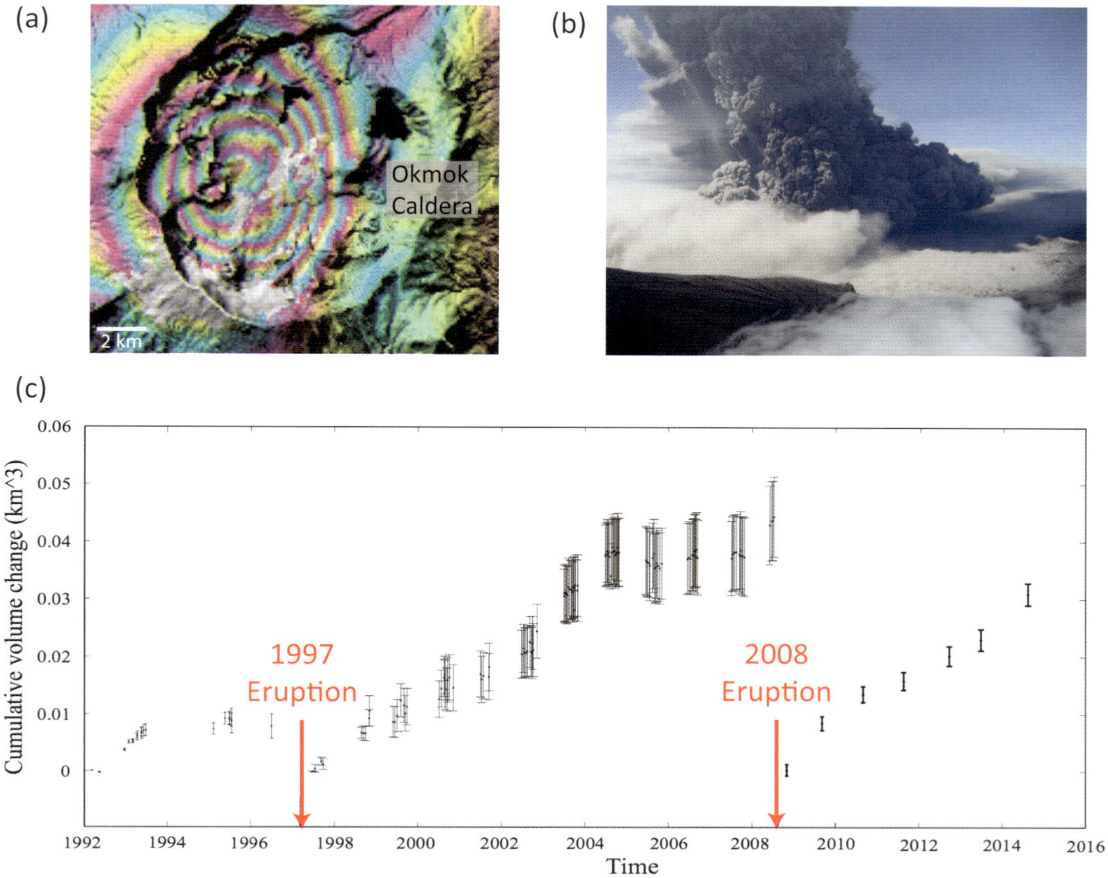

FIGURE 2.3 Some volcanoes display a "breathing cycle" as they inflate, erupt, and deflate. (a) Interferometric synthetic aperture radar (InSAR) image showing inflation of Okmok volcano, Alaska, in 2002–2003, where each interference fringe (complete color spectrum) represents 2.83 cm of change in distance between satellite and ground. (b) 2008 eruption of Okmok. (c) The inflation–deflation cycle at Okmok volcano. The volume change in the magma reservoir is calculated from deformation recorded as in (a). The increase in volume after the 1997 eruption is interpreted as an influx of magma. The ground surface then rapidly subsided during the 2008 eruption, and the reservoir is inflating now again. SOURCE: Modified from Lu and Dzurisin (2014) with new data courtesy of Zhong Lu, Southern Methodist University.

bubble geobarometers that can be used to infer the depth of crystal formation (Aiuppa et al., 2007; Neave et al., 2015).

Despite this large geophysical and geochemical toolkit, resource constraints mean that few volcanic eruptions have been recorded using more than the most basic seismic, deformation, and gas instruments, and both long-term and real-time measurements are often absent or incomplete.

To Ascend or to Stall?

Physical processes and the rheology of the crust govern whether magma ascends from its source to erupt, or stalls and accumulates without erupting. In general, magma will stall if it loses buoyancy, increases in viscosity, or can no longer open and flow through vertical cracks in the surrounding crust. Thus, the intrinsic characteristics of the magma matter: its chemical composition, dissolved and exsolved gas content, temperature, and crystallinity, all of which affect magma density, compressibility, and rheology. Extrinsic parameters also matter, including the density, strength, and stress state of the surrounding crust, and pre-existing weaknesses and structures in the crust. The tools of volcano science are starting to be able to sense magma movement and storage in many regions, and yet it is not known which combination of intrinsic and/or

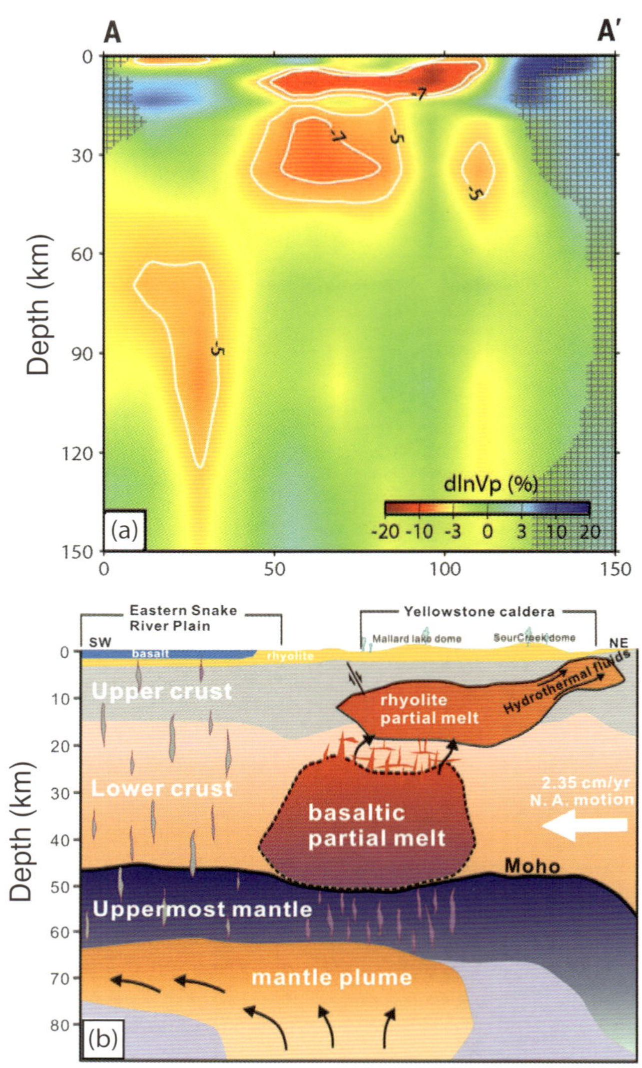

FIGURE 2.4 Seismic imaging shows the location of the Yellowstone magma reservoir in the crust and plume in the mantle. (a) Crustal regions with melt present are illuminated in this tomographic P-wave speed model with low seismic wave velocities (red and orange). This cross section runs through the long axis of Yellowstone caldera. (b) Cartoon interpreting the tomographic model as regions of partial melt and a relatively hot mantle plume. Seismic imaging of magma reservoirs typically shows large regions with small percentages of partial melt (or hydrothermal fluid). SOURCE: Modified from Huang et al. (2015).

extrinsic parameters control where magmas stall and accumulate prior to eruption.

The time scales for magma storage and ascent are only now beginning to be quantified. In some settings, magma can be stored for tens to hundreds of thousands of years in magma reservoirs (e.g., Barboni et al., 2016; Kaiser et al., 2017). Some volcanoes erupt magma that has traversed the entire crust (40 km on average) in a few hours to days (Demouchy et al., 2006). Ascent from crustal magma chambers can take only a few hours (e.g., Castro and Dingwell, 2009) or as little as a few minutes (e.g., Humphreys et al., 2008). Constraining these time scales is critical for improving forecasting.

Most magma transport through the crust takes place through cracks known as dikes. Dike propagation involves coupling between fluid flow, solid deformation, and heat transfer (Rubin, 1995). If the melt cannot flow sufficiently rapidly it will cool, become more viscous, and eventually freeze (Rubin, 1993). The direction of dike growth and the focusing of magma toward or away from a central volcano are controlled by the crustal stress state and therefore can be influenced by magma reservoirs (Buck et al., 2006; Karlstrom et al., 2009), surface loading from volcanic edifices (Muller et al., 2001; Pinel et al., 2010), and large-scale stresses and faults. Whether dikes reach the surface depends on magma chamber overpressure, crustal stress, and density structure (Rivalta et al., 2015). As dikes move upward they push the crust aside, often leading to detectable signals in GPS, tilt, strain, and Inteferometric synthetic aperture radar (InSAR) data (Aoki et al., 1999; Segall et al., 2013; Wright et al., 2006). In the shallow brittle crust, this motion is accompanied by rock breakage, in some cases leading to spectacular propagating swarms of earthquakes that can be used to image the passage of magma (Ebinger et al., 2010; Rubin et al., 1998; Sigmundsson et al., 2015).

Recent studies using dense arrays of seismometers have located small earthquakes in vertical clusters, some as deep as the mantle, which could reflect magma transport in dikes (Figure 2.2). Earthquakes below the brittle–ductile transition may be produced by the high strain rate from dike intrusion (White et al., 2011). In volcanic arcs and thick continental crust, magma usually stalls and accumulates within the crust, cools, mixes with other magma, and chemically evolves before erupting. Geodetic, seismic, and petrologic observations typically point to magma storage at depths between 2 and 7 km (Chaussard and Amelung, 2014). Why is there an apparent "sweet spot" for magma storage? Is this where magma reaches neutral buoyancy and is primed for eruption because it saturates in volatiles, reaches a critical viscosity, or encounters a change in crustal stress or strength (Plank et al., 2013)? Detecting where magmas are stored, how they are distributed in space, their intrinsic properties, and the properties of the surrounding crust will require improved imaging using seismic and electromagnetic tools hand in hand with better laboratory measurement of the geophysical properties of partially molten rocks. Understanding the mechanical properties of the crust and magma are essential to answering this basic question.

How Are Eruptible Bodies Assembled and How Long Do They Persist?

Revealing how eruptible magma accumulates and evolves requires determining the time scales for key processes, including the persistence of magma systems and the time to replenish and pressurize magma bodies. There is considerable debate about the length of time that liquid-dominated magmas exist within the crust compared to the longevity of the magmatic system as a whole. Studies of plutons, magma bodies preserved "frozen" in the geologic record, frequently record hundreds of thousands or millions of years of crystallization (Coleman et al., 2012; Miller et al., 2011). Crystals that erupt from volcanoes also yield radiometric ages that can be as old as thousands to hundreds of thousands of years, demonstrating the extreme longevity of many magma systems (e.g., Cooper, 2015; Kaiser et al., 2017; Peate and Hawkesworth, 2005; Reid, 2003; Schmitt, 2011; Zellmer et al., 2005). Moreover, some volcanoes have long repose periods between large eruptions (tens of thousands to hundreds of thousands of years), indicating that it can take substantial time to develop the conditions needed for eruption (Figure 2.5). For other volcanoes, the time to develop those conditions is shorter (e.g., Allan et al., 2013; Cooper and Kent, 2014; Druitt et al., 2012). For example, while the growth of magma bodies may take hundreds of thousands of years, the time between recharge events and eruption for large caldera-forming eruptions may be less than 100 years (e.g., Druitt et al., 2012). Magma may recharge reservoirs in less than days to months

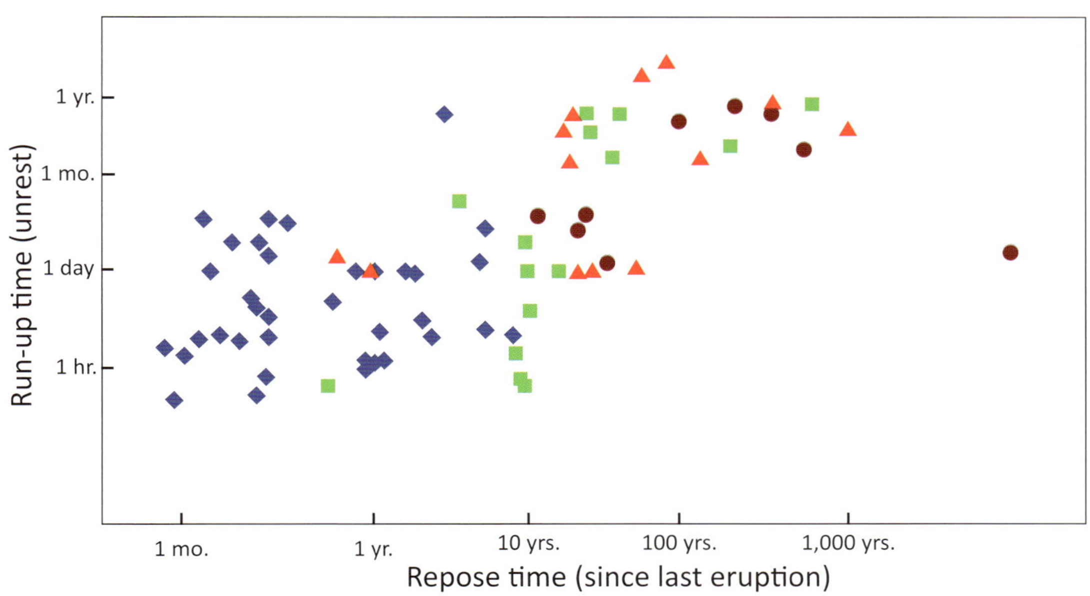

FIGURE 2.5 Recently monitored eruptions show a relationship between repose time, the time between eruptions, and run-up time, the duration of precursory activity. Magma composition is shown with colors (rhyolites, maroon circles; dacites, red triangles; basaltic andesite and andesite, green squares; basalt, blue diamonds). SOURCE: Modified from Passarelli and Brodsky (2012). Reproduced with permission from Luigi Passarelli et al. The Correlation Between Run-Up and Repose Time of Volcanic Eruptions. Geophysical Journal International (2012) 188 (3): 1025-1045. Published by Oxford University Press on behalf of The Royal Astronomical Society (RAS) online at: https://academic.oup.com/gji/article/188/3/1025/683497/The-correlation-between-run-up-and-repose-times-of?searchresult=1. Not covered by any Creative Commons or Open-Access License allowing onward reuse. For permissions please contact journals.permissions@oup.com.

for smaller systems or when magma viscosity is lower (Albert et al., 2016; Rae et al., 2016). Gas emissions often exceed that which can be derived from only the erupted magma (e.g., Christopher et al., 2015; Shinohara, 2008) and suggest degassing and extraction from nonerupted magma remaining at depth. All of these lines of evidence point to the importance of magma accumulation in reservoirs, where it stages prior to eruption, and where a significant proportion of magmas crystallize without erupting. What controls the fraction of magma that eventually erupts? The fraction of magma that erupts is difficult to constrain, exposures of both erupted and unerupted magma are few and far between, and patterns and controls remain difficult to quantify. An upper bound on the long-term average eruption rate is the melt production rate, but in many settings, only a small fraction of magma erupts (White et al., 2006).

The dynamics of magma bodies may be complex and varied. Large and hot magma reservoirs may convect and mix (Bergantz et al., 2015). Thus, the thermal and geometric states of the reservoir are critical to its dynamics and longevity, as they affect viscosity, crystallization, gas exsolution, and freezing (e.g., Gudmundsson, 2012). The primary parameters that control longevity are the temperature of the magma, the mechanical properties surrounding rocks, and the magma flux into the system (Figure 2.6). The

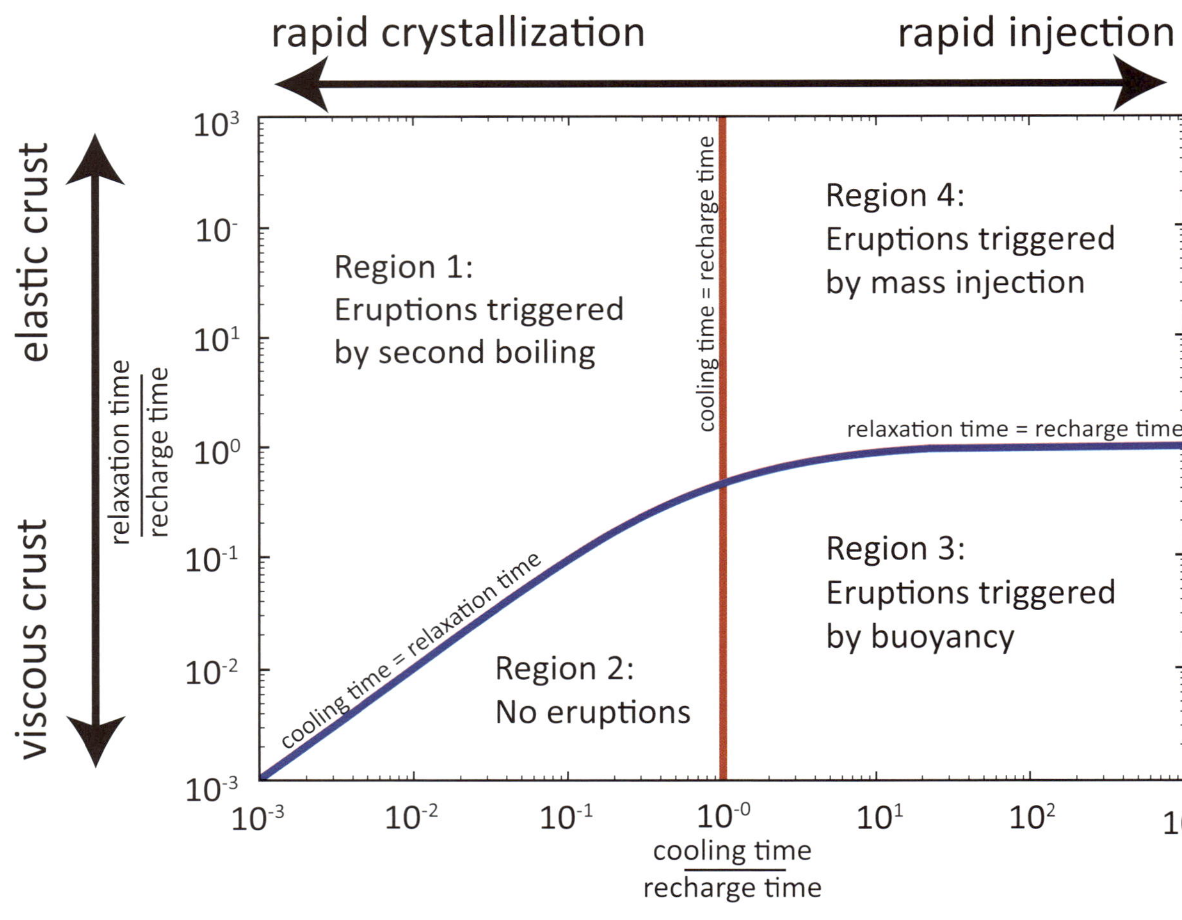

FIGURE 2.6 Conditions that control eruptions for an idealized magma chamber. The horizontal axis is the ratio of heat supplied by new magma input (recharge) to the heat lost by conduction to the surrounding crust. The vertical axis is the ratio of the time scale for viscous relaxation of chamber overpressure compared to the time scale for recharge. Second boiling (renewed vesiculation) refers to the formation of bubbles driven by crystallization. Whether and how an eruption is triggered depends on the interplay of the rate of mass injection into the reservoir, the gas content of the magma, and the size of the reservoir (omitted is the effect of passive degassing). SOURCE: Modified from Degruyter and Huber (2014).

longevity, magnitude, and melt content of eruptible magma bodies have significant implications for hazard assessment and detection of melt prior to eruption and are therefore important targets for future study using combinations of models, geological mapping, geophysical imaging, rock physics measurements, and detailed studies of crystals.

How Quickly Is Magma Mobilized Prior to Eruption?

The regions where magmas stall may help set the course for eruption by influencing the development of overpressure, the accumulation of gas, and the segregation of melt from crystals (Pioli et al., 2009). Magmatic systems that have been in repose for thousands of years may quickly mobilize to eruption following injection of new ("recharge") magma with fresh gas (Bachmann and Bergantz, 2006; Huber et al., 2010). Remobilization might occur several times during transit through the crust, and only the final remobilization may lead to eruption (e.g., Reid and Vazquez, 2017). Evolving conditions during eruption may also mobilize magmas through progressive connection of previously isolated melt lenses (Cashman and Giordano, 2014). New microanalytical techniques (Box 2.2) have recently revealed rich chemical records inside volcanic crystals that may record the timing of injection events days to years before eruption (e.g., Rosen, 2016). Physics-based models predict different triggering mechanisms depending on the magma flux into the reservoir and the behavior of the surrounding crust (Figure 2.6).

Magmatic temperatures and evolution can be constrained using crystal-melt chemical thermometers and diffusion chronometry, but magma recharge volume and history are difficult to constrain. Progress in monitoring magma migration through the crust, accumulation in shallow reservoirs, and the approach to the tipping point for eruption require integrating geophysical measurements, the geochemical and petrologic record preserved in erupted materials, and models that account for the evolution of magma bodies and their interaction with their surroundings.

Key Questions and Research Priorities on Processes That Move and Store Magma

Key Questions

- What causes ascending magma to stall at different levels in the crust and what determines the fraction that eventually erupts?
- Through what processes are eruptible bodies of magma assembled and for how long do they persist?
- How and how quickly do magma bodies mobilize before erupting?

Research and Observation Priorities

- Detect the location of magma by combining geophysical data sets, guided by laboratory measurements of properties of magmas and host rocks.
- Use the chemical composition of minerals and frozen melt to estimate locations and timing of magma residence in the subsurface and ascent to the surface.
- Constrain the intrinsic properties of magma (e.g., density, compressibility, and rheology) over the conditions relevant for eruptions.
- Determine parameters extrinsic to the magma and how these parameters vary over the life cycle of the volcano.
- Build models that couple large-scale transport with processes occurring at the fine scale of bubbles and crystals.
- Integrate the volcanic and plutonic record to elucidate the life cycle of magmatic systems.

2.2 HOW DO ERUPTIONS BEGIN, EVOLVE, AND END?

Anticipating when an eruption will begin, how it will evolve over time, and when and why it will end are among the greatest challenges in volcano science. Most volcanoes are not continuously active but spend much of their lifetime at rest, sometimes for thousands

BOX 2.2
Eruption History Recorded by Crystals and Vesicles

Crystals and bubbles that form below a volcano are carried upward in the magma during volcanic eruptions. These crystals and bubbles serve as chemical and physical archives of the processes operating during storage and transport of magma, and they can be used as geothermometers (temperature), geobarometers (pressure), and geospeedometers (rate of ascent). Information on these processes can be extracted in several ways. First, the number, size, and shape of crystals and bubbles reflect a balance between the nucleation of new objects and their continued growth, which, in turn, constrain variables such as cooling rate, decompression rate, crystallization rate, and the supply rate of raw material for growth. Second, the chemical composition of individual crystals, the assemblage of different minerals that are present in the magma, and the distribution of elements between coexisting minerals and the melt reflect the thermal and physical state of the melt (temperature, pressure, and composition). Crystals and melt may become chemically zoned during changes in magma state. Chemical diffusion then works to smooth these zones, and the extent of this smoothing provides a duration of subsurface processes, which are most usefully interpreted if samples are collected within a well-constrained stratigraphic framework or from dated eruptions. For example, Mg/Fe zonation profiles in crystals were used to date events in the months leading up to the explosive May 1980 eruption of Mount St. Helens (see figure below, Saunders et al., 2012), as one magma mixed with another. The ability to use radioactive isotopes to date crystals adds the dimension of absolute time, which is critical for understanding how quickly magma systems evolve and how changes relate to the monitoring signals recorded at the surface.

These approaches have led to insights on the time scales of magma storage and ascent. For example, preservation of chemical zonation inside crystals that grew in the mantle requires transport from the mantle to the surface (~40 km) in a few months, much faster than previously thought (e.g., Demouchy et al., 2006). Recent zonation (years to decades prior to eruption) in crystals that originally grew tens of thousands to hundreds of thousands of years ago indicates that magma reservoirs have long lifespans (Cooper and Kent, 2014). The short time scales at high temperature recorded by the crystal zonation indicate that magma bodies can thaw rapidly from a crystal mush to an eruptible state. As a result, we should not expect to detect large, mostly liquid bodies of magma beneath most volcanoes, consistent with seismic imaging (Pritchard and Gregg, 2016). Melt-rich reservoirs are more likely to lead to an eruption relative to the general background state.

The number and dimensions of bubbles and crystals reflect processes such as ascent rate during the late stages of eruption. More recently the size and shape of bubbles have been combined with models of diffusion of volatile species and bubble relaxation to infer time scales and deformation rates during the shallowest ascent. Volcanic rocks with the smallest crystals, the greatest density of bubbles, and zonation of the fastest diffusing volatile elements (like H_2O) indicate very rapid final ascent—minutes to ascend from the final storage region at ~2–10 km depth to the surface. Current research using these fastest volcanic clocks is testing whether more rapid ascent results in more explosive eruptions, as expected because fast ascent allows less time for volatiles to escape. These new tools allow the records contained in crystals and bubbles to be exploited in unprecedented ways, and to constrain the wide range of time scales involved in magma storage and ascent.

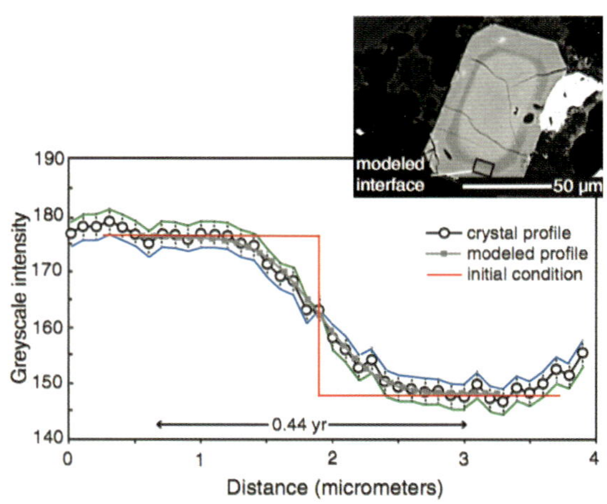

FIGURE Modeled diffusion profile for Mg/Fe ratio (for which greyscale intensity is a proxy) in an orthopyroxene crystal from Mount St. Helens. The profile location is shown in the electron microscope image (upper right). The assumed initial step function in the Fe/Mg ratio in the crystal is shown by the red line; over time diffusion smoothed out this initial step. The time required to reach the measured profile from an initial step function is 0.44 years. SOURCE: Saunders et al. (2012).

of years before erupting again. Prior to an eruption, the movement of magma and fluids may cause earthquakes beneath the volcano, gas emission into the atmosphere or aquifers, or uplift of the ground surface. However, these signs of volcanic unrest do not always presage an eruption. Similarly, some eruptions occur without precursory unrest detectable with our current methods. We still do not know how to interpret the signs of unrest unequivocally. Which are precursors to eruption? What is normal background activity of volcanoes over their life cycle?

Eruptions begin when magma ascends toward the surface, either by propagating in a new dike from the storage region (Section 2.1) or by rising through a preexisting conduit, potentially displacing and interacting with older magma. For volcanic systems in repose, it is commonly assumed that eruptions are preceded by pressure increases within shallow magma reservoirs. The ultimate trigger for eruption can be transfer of additional magma from deeper in the crust (recharge) or changes in the volatile budget (e.g., Girona et al., 2015; Tait et al., 1989). However, the resulting pressure changes may require years, decades, or even centuries to initiate eruption at the surface (Druitt et al., 2012; Morgan et al., 2006). Volcanic unrest that precedes eruptions (run-up) may occur over hours to years (Figure 2.5), although such signals may precede eruptive activity by years to decades (Biggs et al., 2014; Phillipson et al., 2013).

The duration and nature of precursors also depend on the physical setting (tectonic environment and rheology of the crust) and the rheology of the rising magma (e.g., Roman and Cashman, 2006). In general, low-viscosity basaltic magmas ascend rapidly and with brief precursory activity (e.g., Albert et al., 2016; Passarelli and Brodsky, 2012). High-viscosity silicic magmas, in contrast, often have longer run-up periods and may begin with weak gas-driven or phreatic eruptions. This precursory activity acts to construct a magma pathway to the surface (a volcanic conduit). When the conduit is fully developed, buoyancy and the pressure difference between the magma storage region and the surface drive eruptive activity (e.g., Scandone et al., 2007). Some volcanoes, in contrast, maintain connections between magma storage regions and the surface over periods of decades to centuries. These are commonly referred to as open-system volcanoes and may emit gas continuously and erupt with little to no precursory activity. A central challenge is to explain and understand the great variety of styles and durations of all eruptions, and then to incorporate this understanding into eruption forecasting.

Eruption Initiation

Eruptions may initiate either explosively or effusively, and commonly pass from one style to the other during an eruption. Whether exsolved gas escapes from the conduit, favoring lava effusion or causing ascending magmas to stall, or remains physically coupled to the magma, promoting explosive eruption, is strongly influenced by the properties of the melt and the specific nature of the volatile species. Melt viscosity modulates the rates of volatile segregation both prior to and during eruptive activity; for this reason, silicic, high-viscosity magmas are more prone to highly explosive activity than mafic, low-viscosity magmas.

Fragmentation of magma—breakage into small pieces—is a set of critical processes that are required, though not sufficient (Gonnermann and Manga, 2003), for explosive eruption. Rising and decompressing magma will explode, or fragment, if the bubbles contained with the melt cannot expand sufficiently rapidly to accommodate the change in pressure and remain trapped in the melt (Gonnermann, 2015; Proussevitch and Sahagian, 1998; Sparks, 1978; Zhang, 1999), or if strain rates are high enough to drive a viscous magma through the glass transition (Burgisser and Degruyter, 2015; Cashman and Scheu, 2015; Dingwell, 1996; Spieler et al., 2004). Alternatively, rapid decompression and expansion of low-viscosity magma stretches the melt into thin sheets and filaments that are hydrodynamically unstable and tear into fragments (Houghton and Gonnermann, 2008; Namiki and Manga, 2008). Explosive fragmentation can be triggered by sudden decompression during flank failure or collapse of viscous lava flow fronts (Alidibirov and Dingwell, 1996). The former is best exemplified by the May 18, 1980, eruption of Mount St. Helens, when a collapse-triggered lateral blast (Kieffer, 1981; Ongaro et al., 2011) was followed by downward migration of a decompression wave that ultimately intersected the magma storage region and resulted in several hours of sustained eruptive activity (Criswell, 1987; Figure 2.7).

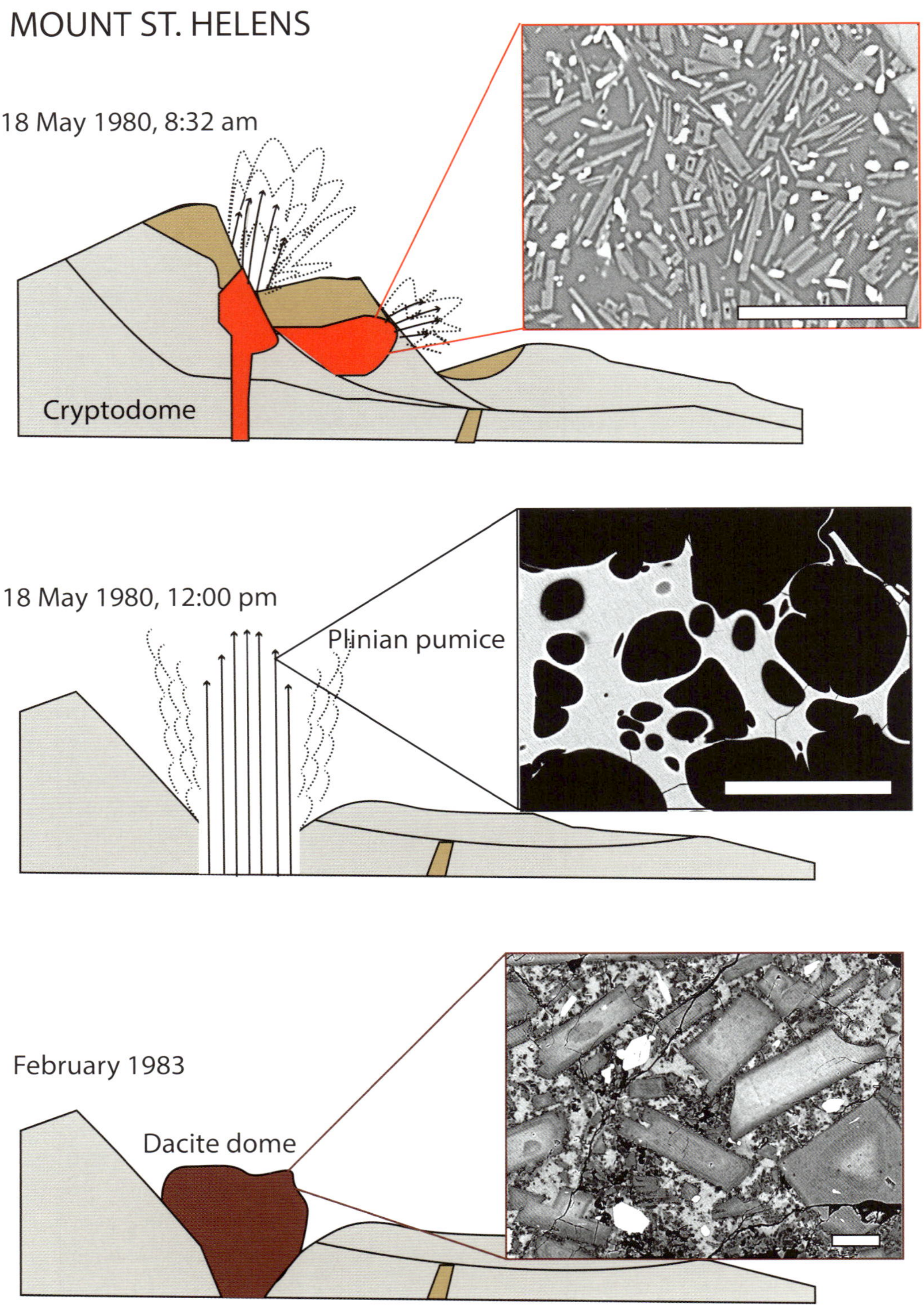

FIGURE 2.7 Evolving eruption style and eruptive products, Mount St. Helens. Schematic cross sections of the volcano from south to north at three periods: (upper) during the initiation of the eruption at 8:32 am on May 18, 1980, by oversteepening and collapse of the north flank because of cryptodome intrusion; (middle) subsequent Plinian eruption lasting approximately 9 hours; and (lower) growth of a dacite lava dome by episodic emplacement of viscous lava flows between June 1980 and October 1986. Insets show backscattered electron images of the microtextures associated with each phase of the eruption. Light gray rectangular crystals are plagioclase, bright white crystals are mafic phases (mostly pyroxene), medium gray is quenched melt (glass), and black holes are vesicles. Scale bar is 25 μm in all images. SOURCE: Redrafted from the U.S. Geological Survey (USGS), http://pubs.usgs.gov/gip/msh/debris.html.

Magma may also fragment nonexplosively, for example by disruption of the crusts of lava flows or in fail–heal cycles during shear deformation of highly viscous magma (e.g., Tuffen et al., 2003). These various fragmentation processes are each fundamentally different, and produce dramatically different particles (Rust and Cashman, 2011).

Conversely, magma will erupt effusively if strain rates remain small enough, if bubbles can expand freely in response to decompression, or if bubbles rise buoyantly or escape through permeable pathways (Gaunt et al., 2014; Jaupart and Allègre, 1991; Okumura et al., 2009; Rust and Cashman, 2004). All of these processes are favored by low-viscosity mafic magmas such as basalt. Nevertheless, there are sustained eruptions of mafic magma (e.g., Vinkler et al., 2012) that challenge this basic understanding.

Many eruptions initiate explosively, often suddenly, even when preceded by weeks to months of precursory unrest. By its nature, the initiation of eruption is difficult to observe in detail, although both seismoacoustic (Johnson and Lees, 2000; Patanè et al., 2013) and radar (Gouhier and Donnadieu, 2010; Scharff et al., 2015) measurements are now being used to characterize the opening seconds to minutes of eruptive activity and the short-term fluctuations in pulsed eruptive activity.

Another common way eruptions begin is steam-driven explosions that occur when magma at high temperature comes into contact with external water (Sheridan and Wohletz, 1983; Zimanowski et al., 2015). Rapid transfer of heat causes water to flash to steam and expand (Kokelaar and Durant, 1983) and magma to quench and fragment (Mastin et al., 2009b; Wohletz et al., 2012). Such phreatomagmatic activity can use thermal energy very efficiently and produce heterogeneous mixtures of juvenile particles, magmatic gas, steam, wall-rock particles, and often, liquid water (Murtagh and White, 2013). Complete mixing is rare and so the products of single explosions may include liquid water droplets and steam as well as both cold and incandescent pyroclasts (Houghton et al., 2015). Open questions remain about (a) the actual mechanism(s) of fragmentation by water interaction as a function of its source (groundwater and surface water) and physical state (ice, liquid, or vapor); and (b) the role of the state of the ascending magma, such as its viscosity and bubble content (e.g., Liu et al., 2015).

When magma ascent is sufficiently slow, eruptions may start effusively. Under these conditions, gas can segregate and outgas from the rising magma at a rate commensurate with that of magma ascent. In fluid magma, this may occur by buoyant rise of large bubbles. In slow-ascending viscous magmas, outgassing occurs through a permeable network of stretched and/or coalesced bubbles and fractures (e.g., Castro et al., 2012; Lavallée et al., 2013). In the former case, eruptions form fluid lava flows. In the latter case, effusion takes the form of thick lava flows or, when decompression also triggers extensive crystallization, viscous lava plugs and domes.

Improved understanding of eruption initiation requires physical and chemical models informed by petrologic, geophysical, geochemical, and observational constraints. As described in Section 2.1, petrology provides a powerful tool for deciphering conditions of magma residence in upper crustal magma reservoirs (e.g., Turner and Costa, 2007). The same petrologic and geochemical tools can be applied to processes in shallow conduits (Rutherford, 2008). For example, micron-scale analysis of crystal-hosted melt inclusions, matrix glasses, and crystals can be used to track the extent of disequilibrium, and hence the time scales, of processes that are responsible for transitions in eruptive activity (e.g., Costa et al., 2003; Humphreys et al., 2008; Lloyd et al., 2014; Watkins et al., 2012; Zellmer et al., 1999). When combined with textural analysis of bubbles and crystals in pyroclasts, these data can be used, in theory, to reconstruct a complete picture of shallow conduit processes (e.g., Cashman and McConnell, 2005; Liu et al., 2015). In a few cases microanalytical data have been linked directly with real-time data from eruptive observations (e.g., Albert et al., 2016; Blundy et al., 2008; Rae et al., 2016; Saunders et al., 2012). These data, however, have yet to be fully integrated into evaluation of precursory activity. Thus, a major challenge is to integrate analytical and experimental data streams into physics-based models for eruption processes that are testable and that can be used to simulate potential eruption scenarios on short time scales.

Eruption Evolution

As eruptions progress, the style and intensity of the volcanic activity are determined, at least initially, by patterns of shallow release (exsolution) and retention or escape (outgassing) of gases initially dissolved in the magma (Burgisser and Degruyter, 2015; Castro and Gardner, 2008). These processes drive much of the rich diversity in eruptive behavior (Gonnermann and Manga, 2012; Figure 2.8). Volatile species have very different solubilities, so that low-solubility CO_2, for example, will start to exsolve at much greater depths than higher-solubility H_2O. Eruption style is also modulated by the depth and geometry of the magma reservoir and volcanic conduit that connects the magma reservoir to the surface. Volcanic products such as lavas and pyroclasts provide our best views of this shallow subsurface magmatic system. A key challenge is to use this erupted material to interpret those subsurface processes that cannot be characterized directly.

The dynamics of eruptive activity can change dramatically with time. For example, initial explosive activity may evolve to short lived (Pinatubo) or long lived (e.g., Kilauea, Santiaguito) effusive eruptions. Alternatively, protracted effusion may be punctuated by larger explosions (e.g., Pallister et al., 2013), and open-system volcanoes may experience rare paroxysmal

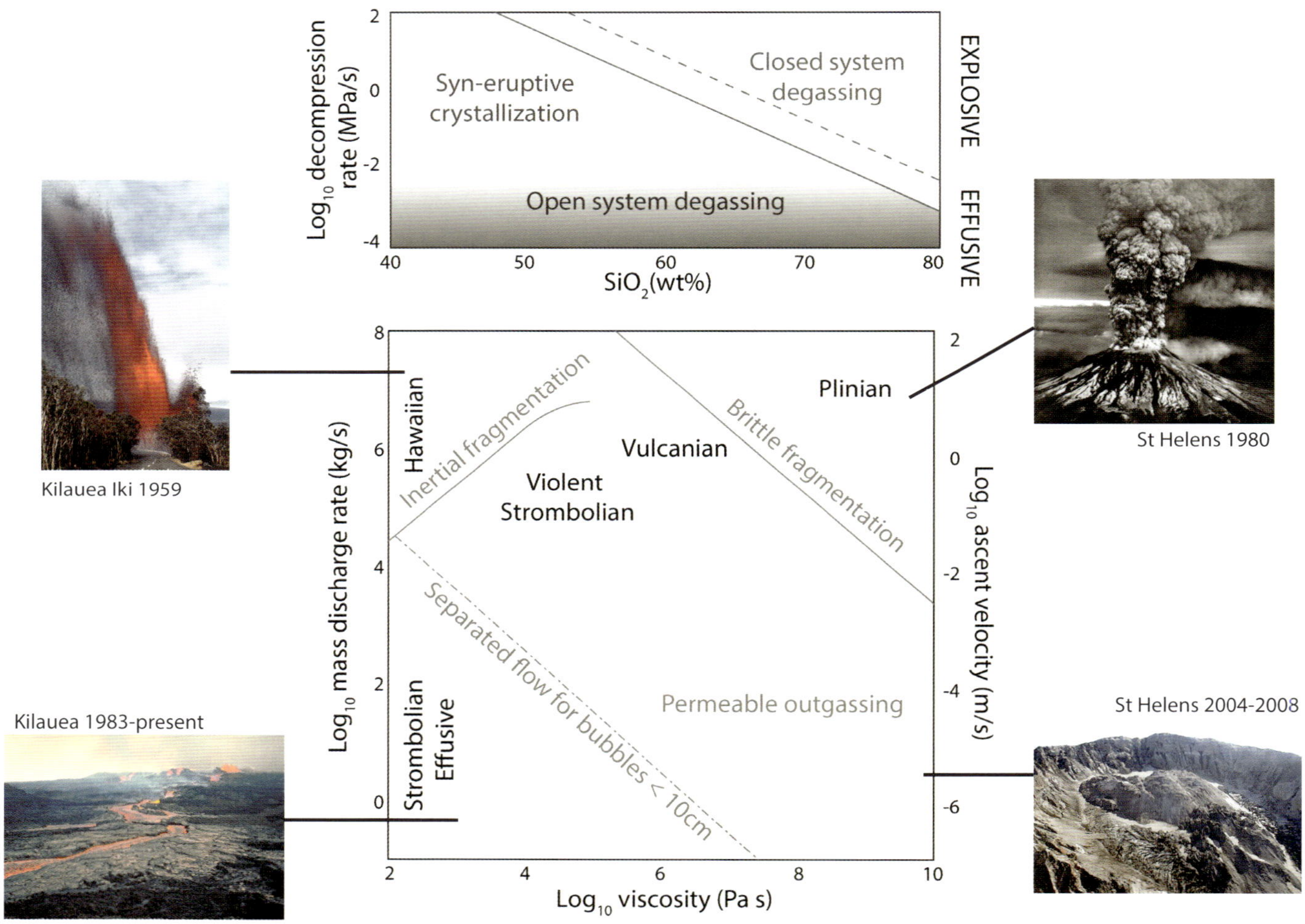

FIGURE 2.8 Regime diagram for eruption style as a function of ascent rate and magma viscosity, assuming steady ascent, based on equations of conservation of mass, momentum, and energy for processes that operate from the bubble scale to the conduit scale (bottom). The gray lines identify critical transitions in the dominance of different processes that govern ascent, including the loss of gas during ascent, developing overpressure in bubbles, crystallization during ascent (top panel), and the ability of magma to fragment. Note that many critical processes are not included in the regime diagram (e.g., unsteadiness, conduit evolution, and interaction with external water). SOURCES: Bottom panel modified from Gonnermann and Manga (2012) with images from the USGS. Upper panel adapted from Cashman and Scheu (2015).

explosions following unloading by lava effusion (e.g., Ripepe et al., 2015) or passive degassing (Girona et al., 2015). Effusive eruption and explosive ash venting may also occur simultaneously (Castro et al., 2012). Thus, a major challenge is to understand both sudden and progressive shifts in activity within eruption episodes. Broadly speaking, shifts in eruptive activity may derive from changes in the source (particularly loss of overpressure from magma discharge, and increases from recharge and exsolution), from changes in the conduit geometry (Michieli Vitturi et al., 2008; Wilson et al., 1980), or from rheological transitions within the magma. Loss of overpressure at the source manifests as an exponential decrease in mass eruption rate, as shown by the 1984 effusive eruption of Mauna Loa volcano, Hawaii, the 1988–1990 effusive eruption of Lonquimay, Chile, and the 2004–2005 effusive eruption of Mount St. Helens, Washington (Figure 2.9). Importantly, all three eruptions showed continuous activity. When activity is discontinuous and accompanied by episodic recharge from depth, the mass eruption rate curves may look quite different, as illustrated by the 1995–2010 eruption of Soufriere Hills volcano, Montserrat. The shape and dimensions of the shallow conduit evolve syn-eruptively by erosion and implosion (Eychenne et al., 2013; Kennedy et al., 2005; Sable et al., 2009). In explosive eruptions, conduit geometry modulates both the eruptive flux and whether the erupted plumes are buoyant or collapse.

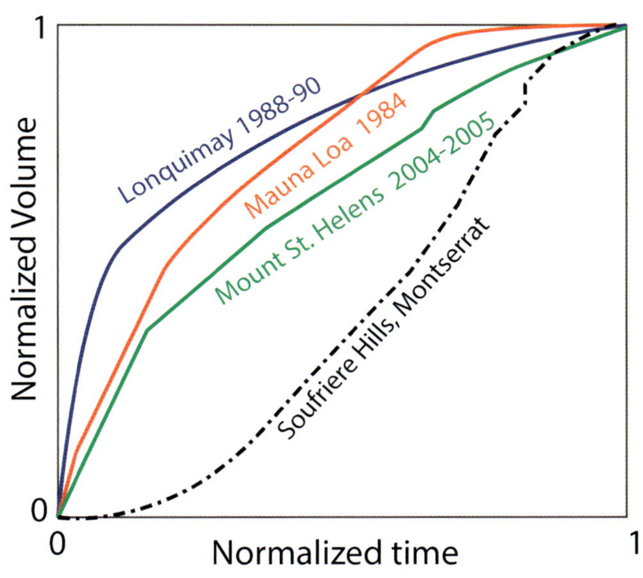

FIGURE 2.9 Effusive eruptions evolve, some showing exponential decreases in volume with time (colored lines). One episode of extrusive activity for Montserrat, in contrast, shows an increase in effusion rate with time, highlighting the great variability of eruption evolution. SOURCES: Data on Lonquimay (1989–1990) from Naranjo et al. (1992), on Mauna Loa (1984) from Lipman and Banks (1987), on Mount St. Helens (2004–2005) from Scott et al. (2008), on Kilauea (1983–1998) from Heliker and Mattox (2003), and on Soufriere Hills (Episode 3, August 2005–March 2007) from Wadge et al. (2014).

Effusive activity is often cyclical (Figure 2.10). Cycles of activity may be generated by elastic deformation (Costa et al., 2007; Maeda, 2000) or stick-slip behavior on conduit walls (Costa et al., 2012; Denlinger and Hoblitt, 1999; Iverson et al., 2006; Ozerov et al., 2003). Cyclical behavior may also be generated internally by nonlinear feedbacks between crystallization and gas loss by permeable flow (e.g., Melnik and Sparks, 2002) or other rheological changes (e.g., Michaut et al., 2013). All of these interactions change with magma ascent rate (Figure 2.8), which controls how bubbles and crystals nucleate and grow, how quickly gas segregates from the melt, how magmas heat frictionally along conduit walls, and how magma may pass from fluid to brittle behavior. Feedbacks between processes are common and include the following:

- Changes in crystallinity can cause magma to cross rheological thresholds, localizing deformation, promoting fragmentation, and changing eruption style
- Changes in gas segregation and gas pressure can cause rapid shifts between degassing regimes and changes in melt rheology
- Changes in heating by friction or crystallization can alter the mechanism of magma ascent
- Transitions in deformation behavior can cause magma to break rather than flow

One critical step for improving our understanding of eruption initiation, modulation, and termination is to quantify the key physical processes in the shallow conduit that are not yet well understood or are poorly constrained by data. These include rheological changes caused by changes in the abundances of bubbles and crystals, interactions among bubbles (including coalescence) and crystals, conditions and rates of permeable degassing, thermal effects of flow and phase transitions, mixing and interaction with host rocks, frictional behavior of magma, and modulation of frag-

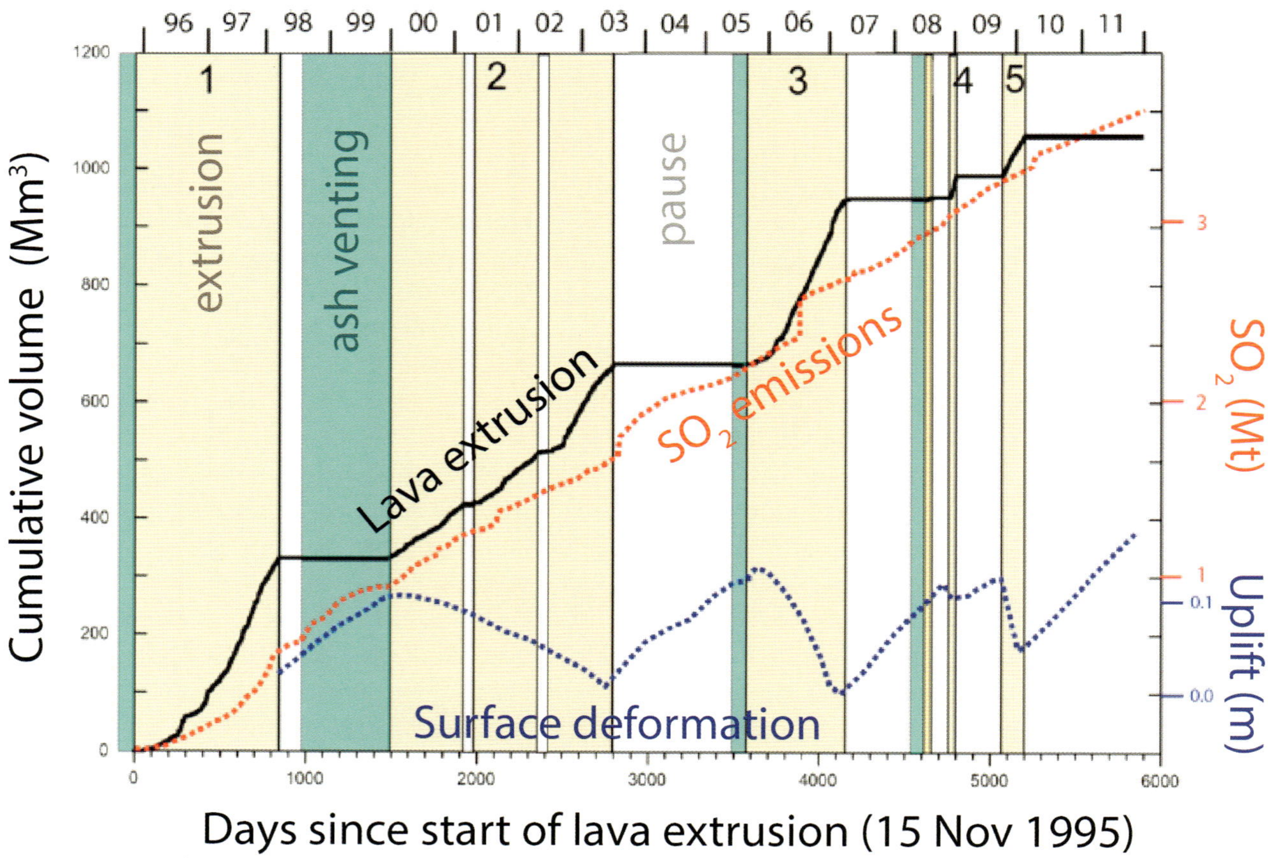

FIGURE 2.10 For 16 years, Montserrat erupted in a complex pattern of activity. Lava extrusion occurred in pulses while SO_2 emissions were more continuous. Over the same time period, surface deformation showed long-term cycles of uplift and subsidence likely caused by processes in the magma reservoir; shorter term variations may originate in the conduit. SOURCE: Modified from Wadge et al. (2014).

mentation and transport processes by interacting with water (i.e., ice, liquid, and vapor). Some of these gaps can be filled by experiments on natural magmas (e.g., Kueppers et al., 2006; Lindoo et al., 2016; Mangan and Sisson, 2000; Okumura et al., 2009; Pistone et al., 2012; Takeuchi et al., 2008) and analog materials (e.g., Castruccio et al., 2013; Cimarelli et al., 2011; Mueller et al., 2011; Oppenheimer et al., 2015; Valentine and White, 2012). Understanding other processes requires dynamic, time-varying models of multiphase flow that couple large-scale transport with processes at the scale of particles. The range of flow behaviors that may arise from nonlinearities in these complex systems is illustrated by the huge oscillations in flow rates and flow regime affected by small perturbations in two-phase flow systems (Melnik and Sparks, 2002; Pioli et al., 2012).

New observational research is also needed to understand controls on the evolution of eruptive activity. In the past few decades, observations of volcanic eruptions have improved dramatically thanks to new satellite-based observations and high-precision geophysical instruments. High-speed visual, thermal, and ultraviolet cameras now permit measurement of key parameters (eruption velocity, mass eruption rate, particle size, and gas flux) on time scales greater than 1 Hz, appropriate for quantifying fine-scale variations in explosive activity (Taddeucci et al., 2012). Effusive activity is well characterized by this new technology, as are Strombolian and, to a lesser extent, Vulcanian explosions. More challenging is acquisition of equivalent high-resolution data sets for sustained explosive eruptions (i.e., subplinian, Plinian, and Hawaiian high-fountaining eruptions). Such events occur less frequently and typically last only hours to days. Plinian explosive eruptions, in particular, produce large, dynamic, and optically opaque plumes. Characterizing them in real time will require rapid-response deployments and direct links to sample collection and deposit-focused studies with fine-scale temporal resolution.

Eruption Termination

One of the most difficult challenges in volcano science is to determine when an eruption is over, especially when it includes multiple episodes and long pauses (Sheldrake et al., 2016). In the simplest case (e.g., effusive eruptions), an eruption may tap a single, isolated pressurized magma chamber, eruptive activity is continuous, the mass eruption rate decreases exponentially with time (Figure 2.9), and the end of the eruption can be anticipated with some degree of accuracy (e.g., Kauahikaua et al., 1996). Often, however, eruptions tap more than one magma storage region (e.g., Tarasewicz et al., 2012), or magma is resupplied to the system between eruptions (Figure 2.10), or the system becomes "open," so that influx balances output (Poland et al., 2014). Under these conditions, eruption terminations are currently impossible to anticipate, yet the answer is important for forecasting, especially when unrest persists long after the eruption. New insights may come, however, from emerging conceptual models of magmatic systems. In particular, by considering the broad range of scales in magmatic systems, from the crystal- and bubble-scale to the scale of magma bodies, it is possible to develop more comprehensive models for long-term patterns of eruptive behavior whereby magma reservoirs at all depths interact with each other (e.g., Christopher et al., 2015).

Ultimately, the evolution and end of a volcanic eruption may be dominated by processes acting in the shallow conduit. These processes often occur under conditions that are far from equilibrium and that are currently poorly constrained by observations, experimental data, or models. Research advances in observational data will come from new high-density monitoring networks and targeted drilling opportunities. Advances in laboratory experiments will come from real-time and in situ measurements at the high temperatures and relevant pressures of magmatic systems.

Key Questions and Research Priorities on How Eruptions Begin, Evolve, and End

Key Questions

- What processes initiate eruptions, and how can they be identified from geophysical and geochemical precursors?
- How do conduit and reservoir geometries evolve over time?
- What are the critical thresholds in processes and physical properties that govern shifts in eruptive behavior?
- Why do volcanoes stop erupting, and how do we recognize when an eruptive episode is over?

Research and Observation Priorities

- Characterize eruptive processes and products in real time at appropriate temporal and spatial resolutions.
- Perform experimental measurements of the thermophysical properties of magmas and those time-varying and disequilibrium processes that cannot be quantified directly in nature.
- Elucidate interactions between magma and external water, including postfragmentation processes.
- Develop dynamic, time-varying models that couple magma ascent and processes at the scale of particles, bubbles, and crystals.
- Create models of the far-from-equilibrium processes that control the beginning, evolution, and ending of eruptions.

2.3 WHAT HAPPENS WHEN VOLCANOES ERUPT?

Volcanic eruptions distribute lava and volcanic particles over Earth's surface, sometimes to distances of thousands of kilometers. In this sense they are unusual among natural hazard events. Impacts range from highly localized, associated with individual lava flows and near-vent processes, to global in scale when giant calderas form in super-eruptions. Understanding transport dynamics and dissemination of volcanic products over this extreme range of scales is necessary not only for responding to volcanic crises, but also for interpreting the record of prehistoric eruptions (preserved on land and in marine and lacustrine sediments and ice cores) and assessing their impact on Earth systems.

The fate of materials erupted both explosively and effusively is studied using several techniques, including real-time observations of active eruptions, detailed documentation of the physical and chemical properties of volcanic deposits, and physics-based modeling. An overarching goal in volcano science is to understand the links between observed or modeled dynamic phenomena and the deposits they leave behind in the geologic record: this includes plumes and their far-flung deposits, more proximal and highly destructive pyroclastic density currents, and the lava domes and flows produced by effusive eruptions.

Explosive Eruptions: Jets, Fountains, Plumes, and Drifting Clouds

Explosive subaerial eruptions form jets and plumes, consisting of volcanic particles and a mixture of volcanic and atmospheric gases. Plumes may rise buoyantly in the atmosphere, sometimes to stratospheric heights (8–17 km or higher), or collapse under their own weight to produce fountains of ejecta or hot ground-hugging pyroclastic density currents that create distal and near-source hazards, respectively (Section 1.6). These processes interact with both the natural and built environment in complex ways. For example, the otherwise cold and benign falling ash particles that are sucked into airplane engines are reheated and melted, and can create hazards to aviation as well as respiratory problems (e.g., Horwell et al., 2015) and building collapse. The extent to which the jet mixture incorporates and heats the surrounding air controls whether an eruption column rises buoyantly or collapses (Figure 2.11). Models of plume behavior can explain first-order relationships between vent conditions and plume height (e.g., Sparks, 1986; Wilson et al., 1978; Woods, 1988) and collapse thresholds (e.g., Wilson et al., 1980).

Aspects of plume behavior that are not currently well understood fall into three categories:

1. The role of evolving vent conditions, including variations in eruption rate (Clarke et al., 2009; Formenti et al., 2003), overpressured jets and shock waves (Ishihara, 1985; Valentine, 1998), and vent erosion (Solovitz et al., 2014);
2. Dynamics of complex plumes including the generation of pyroclastic density currents and secondary plumes from those currents (e.g., Di Muro et al., 2004; Lara, 2009; Figure 2.11) and their contributions to long-range ash transport (e.g., Eychenne et al., 2012); and
3. The effect of small-scale processes, such as temporally varying grain size and density (e.g., Dufek et al., 2012), and thermal and mechanical energy exchange between gases and volcanic particles (Neri and Macedonio, 1996; Stroberg et al., 2010; Valentine and Wohletz, 1989).

Studies assessing these processes are in their infancy, yet they are critical for quantifying controls on mass partitioning under different eruption conditions.

When carried high into the atmosphere in a plume, volcanic particles are sorted by size and density, with the coarsest and/or densest particles (and aggregates of smaller particles) falling out near the vent within the first few hours of an eruption. Satellite measurements and in situ sampling of volcanic plumes reveal that gases (e.g., SO_2), fine ash, and secondary aerosol particles (e.g., sulfate) may reside in the atmosphere for months to years, and can be distributed around the globe (e.g., Figure 1.1; Carn et al., 2016; Mackinnon et al., 1984; Vernier et al., 2016). Processes that can modify the depositional pattern, but are poorly understood, include ash aggregation (e.g., Brown et al., 2012; Rose and Durant, 2011), ice nucleation (e.g., Van Eaton et al., 2015), hydrometeor formation (Durant et al., 2009), development of gravitational instabilities from particle boundary layers (e.g., Carazzo et al., 2015; Manzella et al., 2015), and orographic effects (e.g., Watt et al., 2015). These processes can remove much of the fine ash prematurely from eruption columns and produce distinctive medial deposits such as a secondary increase in deposit thickness. Aggregation and fallout of aggregates of ash, water, and/or ice from eruption columns are also likely responsible for plume electrification and volcanic lightning in a "dirty thunderstorm" (e.g., Behnke

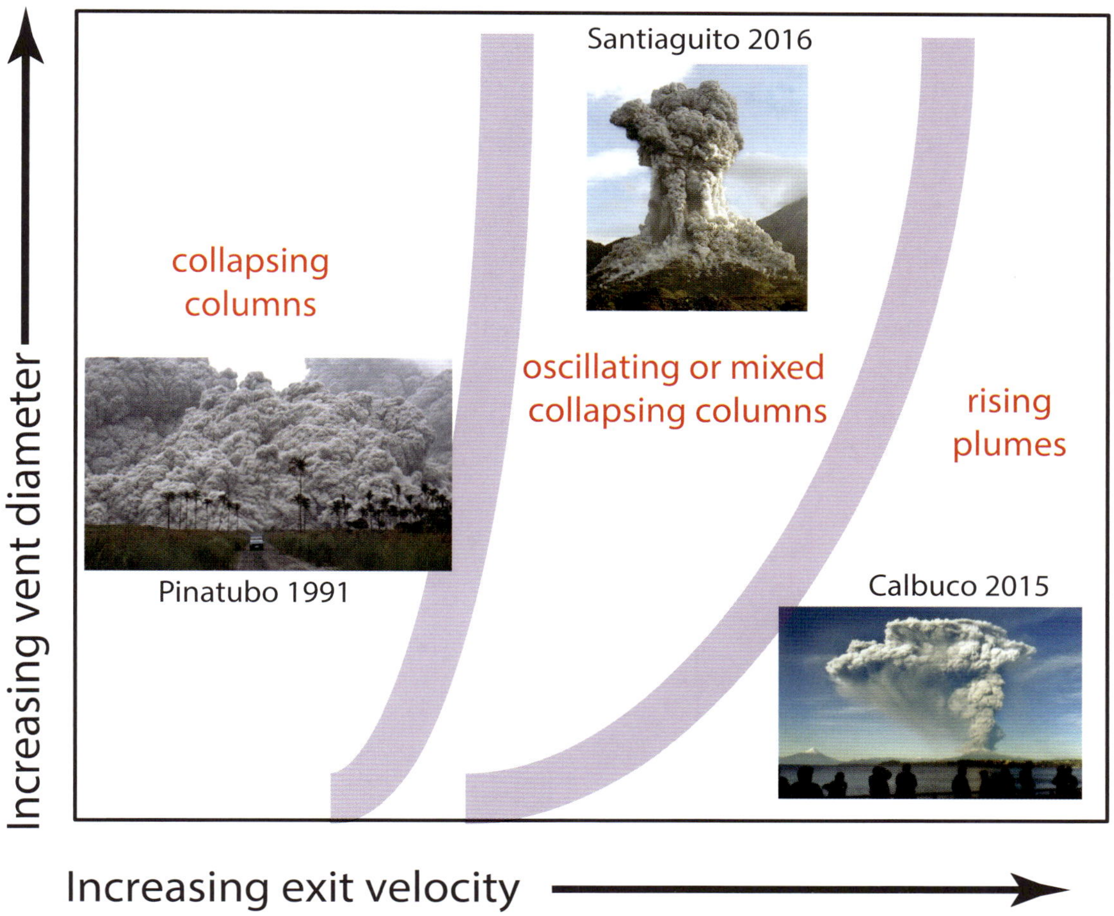

FIGURE 2.11 The fate of volcanic eruption columns depends on the exit velocity and vent diameter, as predicted from multiphase numerical simulations (assuming fixed magma volatile content, steady exit velocity, and two particle sizes). Note the broad zone of oscillating or mixed plume- and pyroclastic density current-forming events, within which behavior may vary greatly. Transitions between characteristic styles (grey lines) shift to the right for increasing particle size and decreasing magma volatile content. Improving the quantitative understanding of the controls on the fate of particles within plumes and the transitions in behavior would improve our ability to forecast the duration and consequences of explosive eruptions. SOURCE: Modified from Neri and Dobran (1994).

et al., 2013; McNutt and Williams, 2010; Van Eaton et al., 2016). Enhanced fine ash deposition reduces ash hazards to aviation and prevents distal ash deposition and preservation in archives such as ice cores. The sequestration of gases by particles in volcanic clouds (Durant et al., 2009) can strongly affect dissemination, residence time, and atmospheric loading of volcanic gases that affect climate (SO_2) and ozone depletion (e.g., Sigmarsson et al., 2013).

Computational models include some of these small- to medium-scale processes (e.g., Oberhuber et al., 1998; Schwaiger et al., 2012; Suzuki and Koyaguchi, 2013), but models are subject to numerous simplifying assumptions (Scollo et al., 2008a,b) and not all processes are sufficiently quantified or even understood. Furthermore, most models of tephra dispersal treat fine volcanic particles and gases as passive tracers in the atmosphere such that the plume itself has no impact on atmospheric temperature and wind patterns, an assumption that may be violated in moderate to large eruptions.

Near-real-time modeling of dispersal processes would also benefit from syn-eruptive measurements of key eruption source parameters. These include eruption onset and end times, changes in the mass eruption rate over time, the total grain size distribution for a range of eruptive styles (e.g., Cashman and

Rust, 2016), the altitude and vertical distribution of gas and ash in an eruption column (e.g., Kristiansen et al., 2015; Mannen, 2014), and particle characteristics such as size distribution, shape, density, and settling velocity (e.g., Alfano et al., 2011; Beckett et al., 2015; Mastin et al., 2009b). Satellite- and ground-based measurements are crucial to determine some of these parameters, either from direct observations or derived from inverse modeling techniques (Eckhardt et al., 2008; Schneider and Hoblitt, 2013). However, future research is needed to develop methods and establish protocols for these difficult measurements. Meteorological parameters, such as wind speed and direction as a function of height and relative humidity, are also required to improve plume dispersion modeling. Research in this area would benefit from strong links to the atmospheric science community and continuous data streams.

Explosive Eruptions: Pyroclastic Density Currents

Field data show that pyroclastic density currents grade from concentrated granular flows to dilute turbulent flows (Branney and Kokelaar, 2002), and often the two occur simultaneously, with a dilute portion overriding a denser basal portion (Valentine, 1987; Figure 2.12). Field studies, scaled experiments, and numerical simulations have been combined to explain depositional and transport processes for a range of different flow regimes (e.g., Breard et al., 2016; Burgisser et al., 2005; Esposti-Ongaro et al., 2012; Roche et al., 2016; Wilson, 1980). Motivating these studies is a set

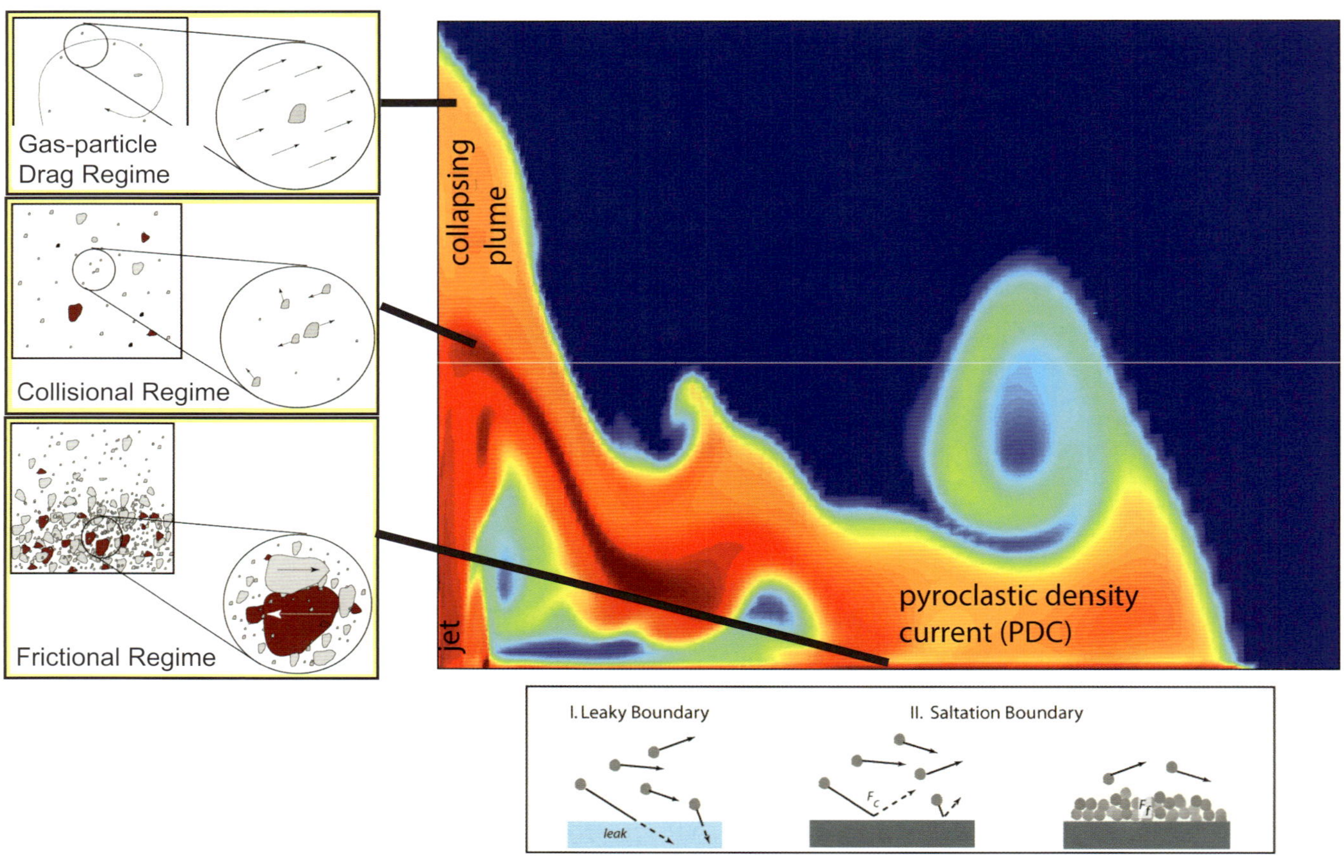

FIGURE 2.12 A collapsing eruption column (main figure) involves several particle-scale processes of mass, momentum, and energy transport. Color indicates particle concentration (red is high, blue is low). The boxes on the left highlight different processes that dominate momentum transport as particle concentration changes. The bottom box illustrates particle-scale processes that represent boundary conditions for large-scale dynamics. Small-scale processes must be understood and modeled correctly in order to capture large-scale, first-order dynamics of pyroclastic density currents and interpret their deposits. SOURCE: Images courtesy of Josef Dufek, Georgia Institute of Technology.

of long-standing and still open questions. How, where, and why does a flow separate into dilute and dense regimes, and what is the corresponding density stratification and mass partitioning? How do these different regimes and partitioning translate into diagnostic deposit characteristics? How should friction in the concentrated granular flow and corresponding erosional power be characterized and quantified? Which types of deposits accumulate principally by aggradation and which are emplaced by en masse stopping? Answers to these questions would inform the preparation of hazard and risk assessments, forecasting areas likely to be impacted and anticipating the consequences of pyroclastic density currents.

To answer these questions, a host of processes must be better understood, characterized, and quantified. Critical small-scale processes include sedimentation (e.g., Bursik and Woods, 1996; Charbonnier and Gertisser, 2008; Komorowski et al., 2013), resuspension (Benage et al., 2016), and particle breakup and comminution (Dufek and Manga, 2008). Such particle-scale processes can lead to order-of-magnitude variability in estimates of runout distances (Fauria et al., 2016) or can even reverse the expected direction of flows (Dufek et al., 2007). Larger-scale processes that require additional research include incorporation of air by entrainment (e.g., Andrews, 2014) and the thermal evolution of the currents (e.g., Caricchi et al., 2014), substrate interaction and erosion (e.g., Brand et al., 2014; Calder et al., 2000; Pollock et al., 2016), and interactions with topography (e.g., Andrews and Manga, 2012; Fisher et al., 1993).

Three approaches will facilitate advances: (1) documenting pyroclastic density current depositional processes in the field, (2) measuring depositional processes in the laboratory, and (3) developing numerical simulations that capture all length and time scales of pyroclastic density current processes (Figure 2.12). The hostile interiors of active pyroclastic density currents have been inaccessible to direct observation; new laboratory-, field-, and drone-based instruments would be transformational in probing these dynamic flows.

Effusive Eruptions: Lava Flows

Effusive eruptions create lava flows and domes. Our understanding of the dynamics of simple, single-lobed lava flows has advanced through a combination of detailed field studies, analog experiments, satellite observations, and numerical modeling (e.g., Harris et al., 2016). However, flows are rarely simple, and quantitative controls on whether a flow will consist of a single lobe or multiple breakout lobes are not confidently defined (Maeno et al., 2016; Figure 2.13). This complexity was highlighted by the limited ability to predict the pattern of the June 27, 2014, lava flow from Kilauea that advanced toward the town of Pahoa, Hawaii.

A number of processes that affect lava flow emplacement need to be quantified, including the rheology of crystal- and bubble-bearing lava that evolves during transport and cooling (e.g., Castruccio et al., 2013; Moitra and Gonnermann, 2015; Sehlke et al., 2014); the effect of unsteady effusion rates on the style and distance of flow propagation (Cappello et al., 2016; Favalli et al., 2009; Tarquini and de'Michieli Vitturi, 2014); the mass partitioning between advance of the flow front, breakout lobes, and inflation (e.g., Poland et al., 2014; Tuffen et al., 2013); and the interaction with a sometimes rapidly evolving topography (e.g., Dietterich and Cashman, 2014; Kubanek et al., 2015; Mattox et al., 1993).

Particularly exciting developments in the study of lava flows are new satellite and airborne remote sensing technologies, such as thermal infrared, lidar, and unmanned aerial vehicles, that can provide high-resolution and high-frequency topographic and thermal data for real lava flows (Cashman et al., 2013; James et al., 2007, 2010; Wadge et al., 2014). The ability to quantify rapidly varying effusion rates would complement measurements of flow dynamics enabled by new imaging technologies.

Historical lava flow eruptions do not exceed tens of cubic kilometers (e.g., 1783 Laki, Iceland; Thordarson and Self, 2003). The geologic record, in contrast, shows that prehistoric flood basalt eruptions have discharged thousands of cubic kilometers, with sequences of these large flows (large igneous provinces) comprising millions of cubic kilometers of lava covering hundreds of thousands of square kilometers (Coffin and Eldholm, 1994). Because we have never witnessed such events, we know little about the conditions of eruption, including both instantaneous and long-term effusion rates (e.g., Self et al., 1997), nor are the geometries of storage reservoirs well understood (e.g., Karlstrom and

Richards, 2011). Addressing our observational bias represents an important challenge, not only to improve our understanding of the dynamics of large-volume events, but also to understand their impact on Earth systems (e.g., Black et al., 2014).

Effusive Eruptions: Lava Domes

Silicic and crystal-rich lava domes are the most viscous type of effusive eruption. Emplacement dynamics have been studied extensively in both the laboratory and during several recent and well-observed eruptions (e.g., Mount St. Helens, Soufriere Hills, Merapi, Santiaguito, Chaiten, and Cordon Caulle). Laboratory experiments have demonstrated that the time scale of lava effusion relative to cooling controls dome and flow morphology, and field studies have shown that the theoretical and experimental framework transfers effectively to effusive eruptions (e.g., Buisson and Merle, 2002; Griffiths and Fink, 1997).

Although effusion rates are typically low and lava flows are typically short in length, domes can suddenly collapse or explode to form pyroclastic density currents and lateral blasts (e.g., the 1997 event at Soufriere Hills volcano, Montserrat; see Belousov et al., 2007; Hoblitt et al., 1981; Sparks and Young, 2002), or vertical eruption columns (Carn and Prata, 2010; Druitt et al., 2002; Robertson et al., 1998). A number of factors can influence collapse, including effusion rate (Calder et al., 2002; Carr et al., 2016; Nakada et al., 1999); dome volume, geometry, or strength (Loughlin et al., 2010; Simmons et al., 2005); permeability and pressurization (Fink and Kieffer, 1993; Voight and Elsworth, 2000); and rainfall (e.g., Carn et al., 2004; Elsworth et al., 2004; Matthews et al., 2002; Taron et al., 2007). However, we still cannot predict the dimensions, style, and timing of such events (e.g., Miller et al., 1998; Watts et al., 2002).

Dome-forming eruptions also tend to be long lived (years to decades) but may be episodic with lengthy pauses in eruption (e.g., Soufriere Hills volcano, Montserrat; Wadge et al., 2010; Figure 2.13). The controls on the tempo of eruption and magma supply (Section 2.2) remain poorly understood. New types of measurements promise to provide critical insights. For example, during extrusion hiatuses, measurements of gas emissions can provide constraints on continued magma supply from depth (e.g., Christopher et al., 2010). Sudden transitions from effusive to explosive activity in these long-lived eruptions remain among the most challenging characteristics to explain and forecast.

Secondary Processes: Lahars

The products of eruptions are subject to a range of secondary processes often operating on far longer time scales than the parent eruptions (Major et al., 2000). Principal among these are volcanic mudflows (lahars) and floods produced when large masses of water mix with volcanic sediment and sweep down the slopes of volcanoes, incorporating additional water and sediment (Vallance and Iverson, 2015). The effects of lahars and floods often extend well outside the primary footprint of eruptions. For example, the 1985 eruption of Nevado del Ruiz, Columbia, was relatively small (Volcano Explosivity Index [VEI] 3), but it generated a syn-eruptive lahar that was 10 times larger in volume and traveled up to 100 km, killing more than 23,000 people (Pierson et al., 1990). The VEI 6 Pinatubo eruption in 1991 was followed by a decade of devastating floods and lahars extending in space and time well beyond the pyroclastic density current deposits that spawned them (Rodolfo et al., 1996).

Lahars and floods share a number of common transport and deposition processes with pyroclastic density currents. However, the complex rheology of lahars is unusual in the range and extent of downstream flow transformations produced by the competing effects of dilution (addition of water), bulking (erosion of sediment), deposition, and infiltration of water into the substrate. No single lahar can be uniquely assigned a flow state that is applicable over its entire depth range and lifespan, yet this assumption is frequently adopted for models and hazard assessments. The timing of lahar events is largely unpredictable at present, and models for their flow do not have satisfactory equations to describe the evolution of flow density and bed erosion with time and distance (Vallance and Iverson, 2015).

Water–Magma Interactions

Magma ascending through the crust often interacts with external water, such as groundwater, lakes, oceans, and ice. At one extreme, phreatic eruptions

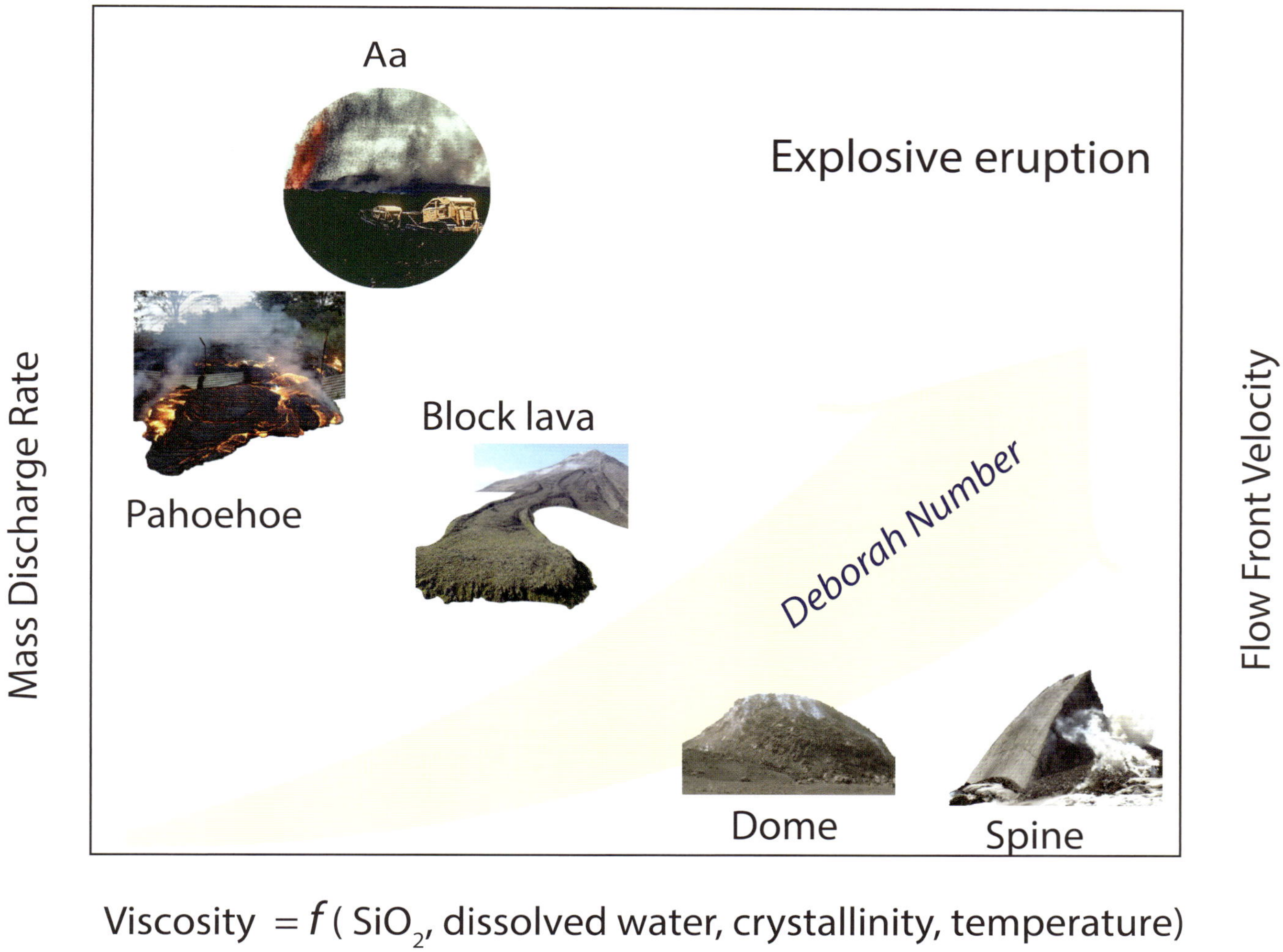

FIGURE 2.13 Effusing magma will form simple flows and domes whose length and form depend on magma viscosity and effusion rate. Spines are a high-viscosity and low-effusion rate end member. The Deborah Number (*De*) is a dimensionless number that characterizes the fluidity. Lava is more fluid at low *De*, and demonstrates solid-like behavior at high *De*. Transitions are not well constrained and are defined by complex and rapidly varying lava characteristics such as crust yield strength and thermal diffusivity (Blake and Bruno, 2000), creating a challenge for quantifying the transitions. Aa (upper left): 1960 eruption of Kapoho, Hawaii, which destroyed 400 buildings. Pahoehoe (left): Narrow flow tongue invading Pahoa orchard, Hawaii, on October 28, 2014. Block lava (center): 1960 eruption of Kanaga volcano, Alaska. Dome (right): Mount St. Helens, Washington, dome on August 22, 1981. Spine (far right): Mount St. Helens on April 26, 2006. SOURCES: Concept from Maeno et al. (2016), which follows from Walker (1971). Images courtesy of the USGS.

occur when groundwater flashes to vapor upon contact with hot rock or magma, but no juvenile magma is erupted. Recent phreatic activity at Te Maari craters, New Zealand (Breard et al., 2014), and Ontake and Aso volcanoes, Japan (Kaneko et al., 2016; Kato et al., 2015), highlight the hazard of these events, which can be highly explosive and often occur without apparent warning. Phreatic events are often interpreted as critical precursors to magmatic eruptions, although they may also occur in isolation. Eruptions driven primarily by the explosive interaction between magma and water are termed phreatomagmatic (Morrissey et al., 2000). Such eruptions are characterized by violent explosions, volcanic plumes, ejection of large ballistic blocks, dilute pyroclastic density currents that spread radially (pyroclastic surges), and lahars (White and Houghton, 2000), and the resulting landforms include tuff rings, tuff cones, and maars (White and Ross, 2011). Ash generated by phreatomagmatic eruptions tends to be finer grained than ash from purely magmatic explosive eruptions due to highly efficient fragmentation (Walker, 1973). As a result, ash will stay in the atmospheric longer unless counteracted by enhanced ash aggregation and premature deposition in wet eruption plumes (e.g., Brown et al., 2012). The eruptions themselves tend to be unsteady, often pulsating at high frequency, and they can be highly destructive, since thousands of pyroclastic density currents can be generated during a single eruptive episode (Brand and Clarke, 2009). Under the right conditions, magma–water mixing produces repeated explosive bursts caused by rapidly expanding water vapor along with magma quench and fragmentation, a process distinct from, although possibly aided by, purely magmatic fragmentation (Büttner et al., 1999; Zimanowski and Büttner, 2003). Quantitative advances require experimental, numerical, and field studies focused on the coupled mixing and fragmentation processes, fine ash formation, and the resulting style, scale, and duration of eruption.

Submarine eruptions represent another extreme end member of magma–water interaction. Such eruptions represent 75 to 80 percent of all magma erupted on Earth, with basaltic magma erupting at mid-ocean ridges to form the oceanic crust and at intraplate hot spots to form ocean islands and seamounts, and more silicic magmas erupting at submarine volcanic arcs. The hydrothermal systems overlying submarine volcanoes can reach very high temperatures due to the high hydrostatic pressure, and their fluids support unique chemosynthetic ecosystems (e.g., hot vents, cold seeps, mud volcanoes, and sulfidic brine pools) and concentrate valuable metal ores.

Submarine eruptions are predominantly effusive along mid-ocean ridges and produce a range of forms, including pillow basalt flows, broad thin sheets, or domes. Due to their inaccessibility, they are understudied relative to their counterparts on land. However, hydrophone networks and a new cabled observatory on the Juan de Fuca ridge (Barnes et al., 2007), have increased our capacity to detect effusive eruptions at mid-ocean ridges and to study their products using deep sea robotic and manned submersibles (e.g., Chadwick et al., 2016; Rubin et al., 2012; Soule et al., 2007; Wilcock et al., 2016). Given the simplicity of mid-ocean ridge volcanoes (e.g., known magma supply, known crustal thickness, and simple tectonic stress field), understanding how these volcanoes work may be a more tractable problem than understanding their subaerial counterparts. To do so, however, requires a better understanding of eruption sizes and frequencies.

When formed by volatile-rich subduction-zone magmas, submarine eruptions can be highly explosive (e.g., Fiske, 1963; Moore, 1967). Explosive eruptions can initiate with either magmatic and phreatomagmatic fragmentation and range from Strombolian scale to caldera forming (Cas and Giordano, 2014). Recent small-scale explosive eruptions have been observed in the western Pacific using remotely operated vehicles (NW Rota-1, Chadwick et al., 2008; West Mata, Resing et al., 2011). Although they rarely pose a direct threat to human populations, explosive subaqueous eruptions may breach the ocean surface (e.g., 1952–1953 eruption of Myojinsho, Fiske et al., 1998; and the 3 km^3 eruption of Havre in 2012, Jutzeler et al., 2014), and produce large pumice rafts that can adversely affect shipping.

Two central questions about submarine explosive eruptions remain. First, how does water depth affect explosivity? Second, how do wind, ocean currents, and particle settling properties affect geochemical fluxes, pyroclast dispersal, deposit characteristics, and their postdepositional reworking? Theoretical and experimental studies on the interaction between seawater and magma, from the particle to the eruption-column scale, can address these questions about explosivity, transport, and deposition. Monitoring of submarine volcanoes, repeat high-resolution bathymetric surveys with autonomous vehicles, sampling submarine deposits with human-occupied and remotely operated vehicles, and ocean drilling would expand our understanding of the history and nature of submarine volcanism.

Key Questions and Research Priorities on What Happens When Volcanoes Erupt

Key Questions

- How do time-varying mass flux and magma properties affect eruptions?
- How do particle-scale processes influence the large-scale dynamics of plumes, pyroclastic density currents, lava flows, and lahars?
- What is the role of evolving substrates in controlling flow dynamics?
- How do interactions between volcanic plumes and the atmosphere affect the transport and deposition of gas and tephra?
- What processes disperse the products from explosive eruptions driven by magma–water interaction?
- What processes govern the occurrence and dynamics of submarine explosive eruptions?

Research and Observation Priorities

- Develop techniques to measure temporal variations in eruption rate and correlate those to variability in eruptive products.
- Integrate models of small-scale processes and large-scale dynamics of eruptive phenomena.
- Undertake model verification, validation, and comparison.
- Conduct rapid syn- or posteruption collection of critical data to test and improve models.
- Overcome our biased understanding of the spectrum of volcanic eruptions by expanding the study of submarine volcanism, and both small and very large eruptions.

2.4 A COMMUNITY CHALLENGE: MODELING VOLCANIC PROCESSES

A common theme in many of the questions and priorities in this chapter is the importance of developing models to interpret the new generation of high-resolution observations and to enhance understanding of magmatic and volcanic processes. Community-wide model intercomparison and validation exercises can lead to important advances and also highlight deficiencies that need to be addressed by future research. Equally useful is validating models with controlled laboratory experiments and well-constrained field data sets. Two examples in volcano science include a conduit model comparison study (Sahagian, 2005) and an intercomparison of plume models (Costa et al., 2016). Such exercises are particularly valuable when combined with suites of data from laboratory experiments, observations of the geologic record, and targeted real-world case studies.

The largest-volume explosive eruptions have yet to be characterized quantitatively. It remains uncertain how effectively, if at all, our observations of volcanic plumes and pyroclastic density currents from relatively small eruptions scale up to very large eruptions. For example, the rate and processes of radial spreading of large plumes in the atmosphere, both primary plumes and secondary plumes, may vary with the scale of the eruption (e.g., Baines and Sparks, 2005), and the roles of pulsating activity in the largest volcanic eruptions are uncertain (e.g., Self et al., 1984). Numerical models of explosive eruptions provide the means to assess the consequences of yet undocumented eruptions. The combination of modeling and observations provides the basis to overcome the biased understanding of the full spectrum of magmatic and volcanic behavior on Earth.

2
News

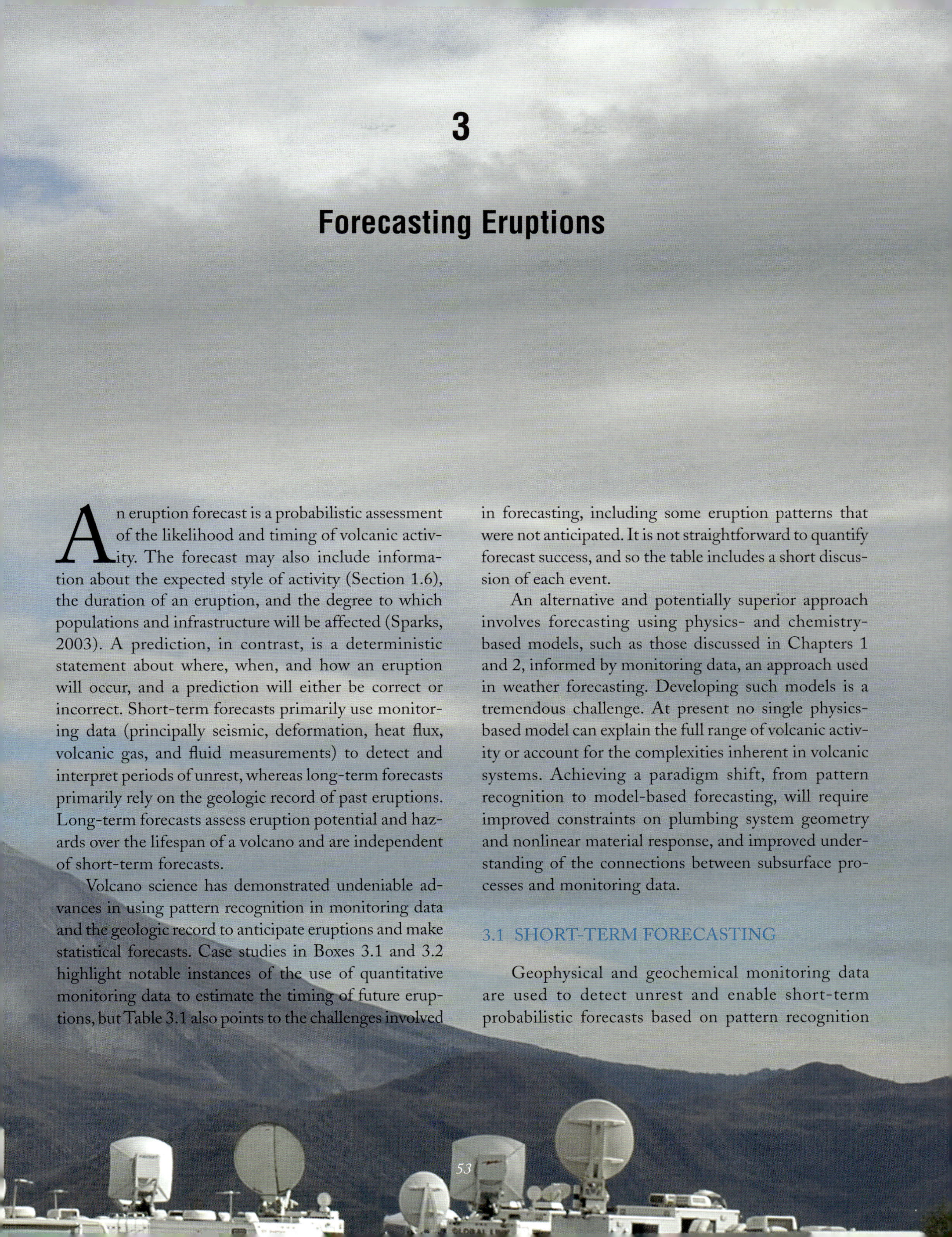

3

Forecasting Eruptions

An eruption forecast is a probabilistic assessment of the likelihood and timing of volcanic activity. The forecast may also include information about the expected style of activity (Section 1.6), the duration of an eruption, and the degree to which populations and infrastructure will be affected (Sparks, 2003). A prediction, in contrast, is a deterministic statement about where, when, and how an eruption will occur, and a prediction will either be correct or incorrect. Short-term forecasts primarily use monitoring data (principally seismic, deformation, heat flux, volcanic gas, and fluid measurements) to detect and interpret periods of unrest, whereas long-term forecasts primarily rely on the geologic record of past eruptions. Long-term forecasts assess eruption potential and hazards over the lifespan of a volcano and are independent of short-term forecasts.

Volcano science has demonstrated undeniable advances in using pattern recognition in monitoring data and the geologic record to anticipate eruptions and make statistical forecasts. Case studies in Boxes 3.1 and 3.2 highlight notable instances of the use of quantitative monitoring data to estimate the timing of future eruptions, but Table 3.1 also points to the challenges involved in forecasting, including some eruption patterns that were not anticipated. It is not straightforward to quantify forecast success, and so the table includes a short discussion of each event.

An alternative and potentially superior approach involves forecasting using physics- and chemistry-based models, such as those discussed in Chapters 1 and 2, informed by monitoring data, an approach used in weather forecasting. Developing such models is a tremendous challenge. At present no single physics-based model can explain the full range of volcanic activity or account for the complexities inherent in volcanic systems. Achieving a paradigm shift, from pattern recognition to model-based forecasting, will require improved constraints on plumbing system geometry and nonlinear material response, and improved understanding of the connections between subsurface processes and monitoring data.

3.1 SHORT-TERM FORECASTING

Geophysical and geochemical monitoring data are used to detect unrest and enable short-term probabilistic forecasts based on pattern recognition

BOX 3.1
Anticipating a Large Eruption Under Challenging Conditions

The 1991 forecast at Mount Pinatubo in the Philippines is widely regarded as one of the greatest eruption forecasting successes in history, saving tens of thousands of lives and millions of dollars in property. Pinatubo, which had not erupted in centuries, was completely unmonitored until a group of nuns living high on the volcano's slopes descended to report that residents of local communities had been feeling earthquakes. Filipino scientists surveyed the volcano by helicopter, noted new gas vents, and installed a single seismometer, which immediately registered a high rate of seismic activity. Recognizing the threat the volcano posed, Filipino scientists invited a small team of U.S. Geological Survey (USGS) scientists from the Volcano Disaster Assistance Program to visit and advise. The USGS team installed additional monitoring equipment and studied previous eruption deposits, quickly amassing strong evidence that previous eruptions were infrequent but large. Over the next 2 months, unrest continued to strengthen, and a sequence of modest but violent "throat clearing" eruptions began. A hazard map was rapidly constructed using maps of large-volume pyroclastic flow deposits on the slopes of the volcano. The map prompted progressive evacuation of a region up to a 40-km radius, only days before the volcano produced the second largest eruption of the 20th century. Due to the evacuation, there were relatively few fatalities given the magnitude of the eruption and the at-risk population. The region was subsequently plagued by years of posteruption lahars and floods often triggered by monsoon rains, resulting in the need for long-term monitoring and mitigation of these secondary hazards. Overall, this eruption was a model for coordinated international rapid response and the integration of geologic and geophysical data to forecast eruption timing and magnitude. It also illustrated the difficulties posed by nonlinear behavior during the run-up to eruption (see figure).

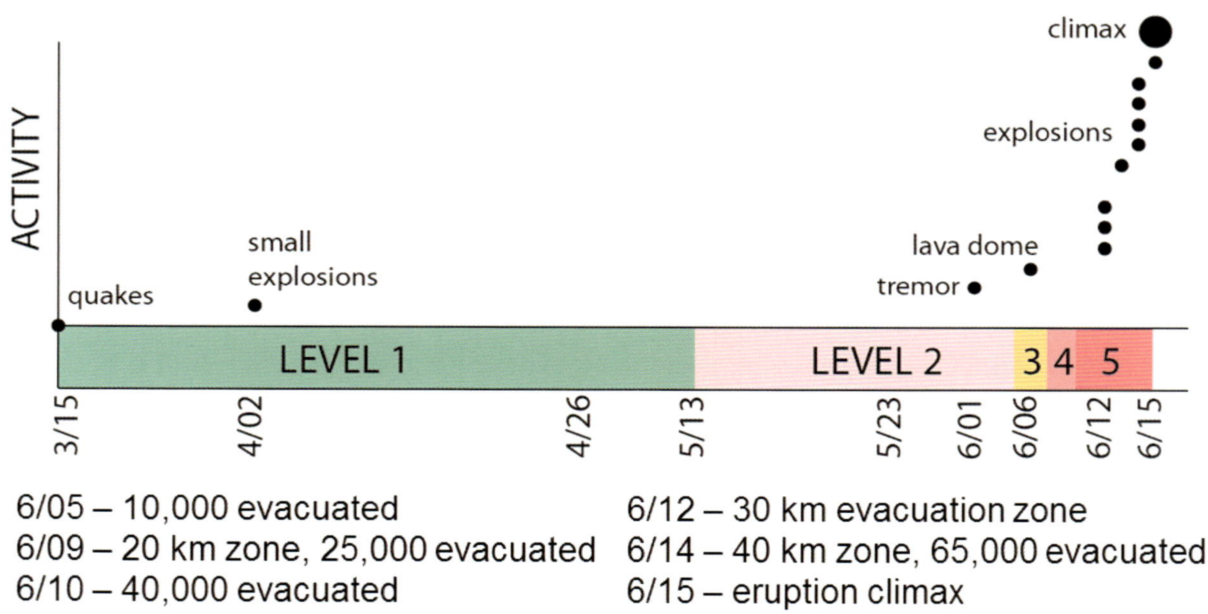

FIGURE Eruption and evacuation chronology leading up to the climactic eruption of Pinatubo, Philippines, in 1991. SOURCE: USGS.

in monitoring time series. As an example, the Mount St. Helens post–May 18 eruptions in 1980 were successfully forecast based on patterns of precursory seismic tremor and localized deformation that consistently preceded events (Malone et al., 1981; Swanson et al., 1983). Relatively frequent eruptions of Kilauea volcano, Hawaii, have led to clear seismic and deformation precursors prior to eruptions. However, many other volcanoes, particularly those that erupt violently, have had limited or no historical eruption observations and few quantitative measurements. In these cases, observations from similar volcanoes are

BOX 3.2
A Precise Eruption Forecast

Undoubtedly the most accurate eruption forecast ever made took place in Iceland in February 2000, when the volcano Hekla erupted following a short sequence of small earthquakes, volcanic tremor, and changes recorded on a borehole strainmeter. The previous eruption of Hekla in 1991 had been preceded by small (M <1) earthquakes and strain signals starting roughly 30 minutes prior to the eruption (Linde et al., 1993). The 2000 eruption followed the previous pattern in remarkable detail (see figure). The first earthquakes were recorded at 17:07 on February 26, 2000. At 17:20 the local volcano science community was notified of a potential eruption, around the time volcanic tremor was recorded at the closest seismometer. At 17:30, shortly after the same strainmeter began showing a signal nearly identical to that of the 1991 eruption, a warning was issued to the National Civil Protection Agency (Stefánsson, 2011). At 17:53, an eruption was forecast to begin in 20–30 minutes, based on the seismicity and strain. A warning was issued on national radio that an eruption of Hekla would occur within 15 minutes: 17 minutes later it erupted. Key for this forecast was instrumentation, the close similarity to past behavior at the volcano, and a record long enough to identify and interpret precursors.

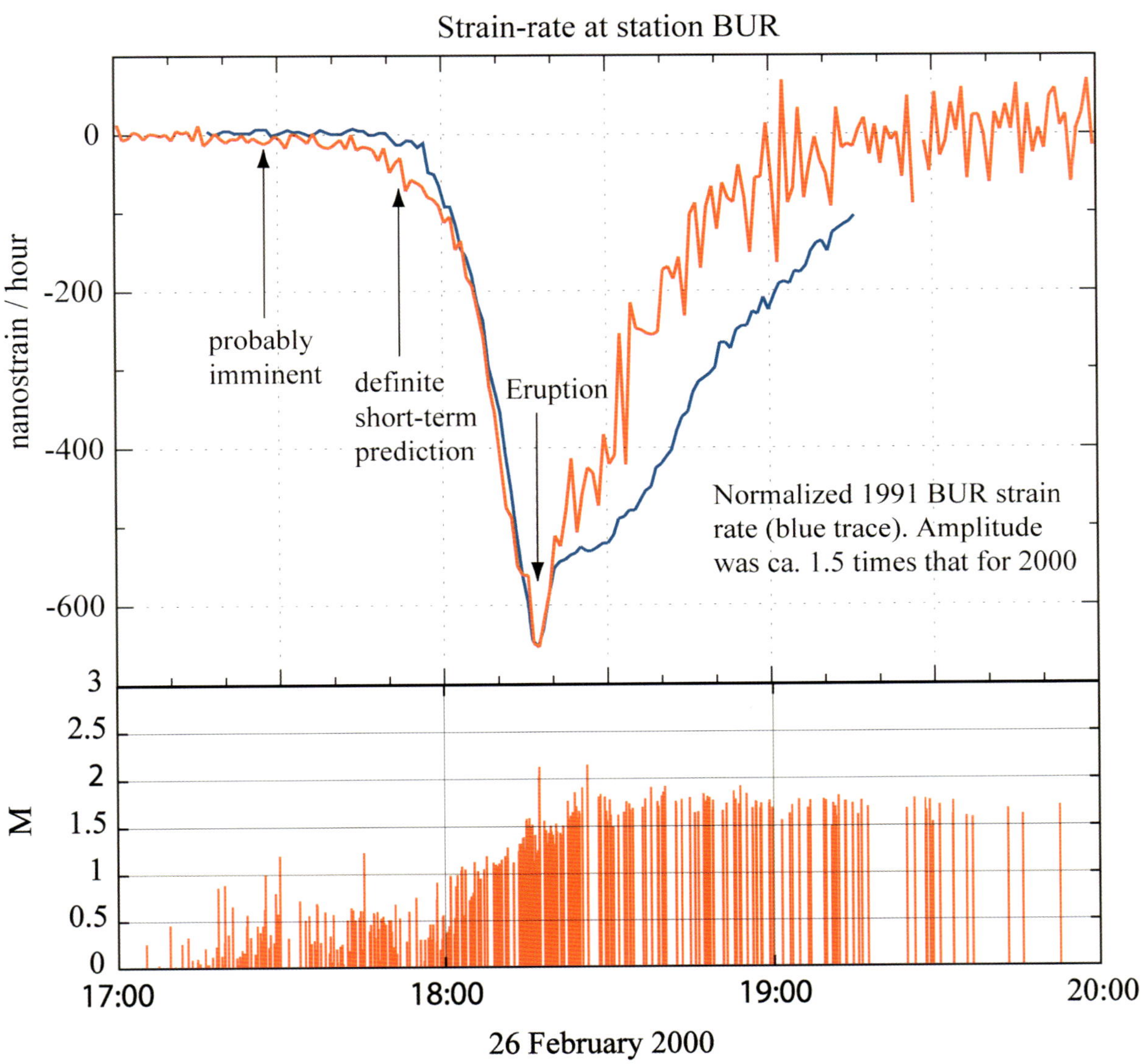

FIGURE Strain rate (top, red line) at station BUR and earthquake magnitudes (bottom) associated with the February 2000 eruption of Hekla volcano, Iceland. Changes in the strain rate and seismicity led to a short-term forecast of an impending eruption, which occurred 24 minutes later. The strain precursor was very similar to that observed prior to the 1991 Hekla eruption (blue line). SOURCE: Stefánsson (2011).

TABLE 3.1 Examples of Forecasts and Missed Opportunities

Volcano	Activity	Successful Forecast?
Mount St. Helens (United States), 1980	Lateral blast and VEI 5 eruption occurred after a period of elevated seismicity, dramatic dome growth, and phreatic explosions (Lipman and Mullineaux, 1981).	Yes and no. Although unrest was observed for months and led to heightened surveillance, the timing, directionality, and scale of the eruption was not anticipated and 57 individuals perished.
Kilauea (United States), 1983–?	Longest eruption sequence (currently ongoing) in Hawaii's history corresponds to the principal locus of eruptive activity ~15 km from the summit on the East Rift. Repeated episodes of lava fountaining and effusion were well monitored. The geodetic network at the summit is used to anticipate downrift eruptive activity with hours of warning (Anderson et al., 2015).	Generally yes as the eruption progressed. Observations led to a viable model, which was used to estimate a high likelihood of events during certain time intervals. The long duration of activity (more than 30 years) was not anticipated.
Nevado del Ruiz (Colombia), 1985	Modest eruption spawned large lahars, which was a hazard previously recognized in both historical and geologic observations. Phreatic eruptive activity and elevated volcanic gases pointed to unrest beginning 1 year prior to eruption (Pierson et al., 1990).	No. Although elevated activity spurred volcano study and the development of hazard maps, government agencies were unable to forecast the primary hazard and provide guidance in a timely manner. More than 25,000 deaths resulted.
Pinatubo (Philippines), 1991; see also Box 3.1	Largest eruption of the last hundred years. Its potential Plinian eruption size was anticipated based on studies of previous eruption deposits. Eruption evolution was relatively "well behaved," with seismic precursors, phreatic explosions, tiltmeter inflation, increasing sulfur output, then increasingly violent magmatic eruptions (Newhall et al., 1996).	Yes, in the sense that evacuations (out to 40 km) were issued prior to the paroxysmal event and valuable property at Clark Air Force Base was moved in a timely manner. Still, some 800 people were killed, largely due to roof collapse and lahars.
Soufriere Hills (Montserrat), 1995–?	Long-term small to moderate ash eruptions beginning in 1995 were later accompanied by lava-dome growth and pyroclastic flows that forced evacuation of the southern half of the island and ultimately destroyed the capital city of Plymouth, causing major social and economic disruption. To date, there have been four phases of eruption, separated by periods of up to 2 years with no residual surface activity.	Yes and no. Cyclic short-term (6–12 hour) precursors, such as increased seismicity and inflation, were successfully used to anticipate the most dangerous times and the most likely timing of dome-collapse pyroclastic density currents. However, the long-term behavior and ultimate duration of the eruption have been difficult to anticipate.
Hekla (Iceland), 2000; see also Box 3.2	An eruption producing an 11-km-high ash plume was accurately forecast (Höskuldsson et al., 2007). Timely notifications were made to the National Civil Defense of Iceland within 20 minutes of unrest and 40 minutes prior to the inferred onset of eruption. The Civil Aviation Administration was also notified.	Yes. A well-monitored volcano presented precursory earthquake activity that was well understood and similar to a previous eruptive episode. The data were used to anticipate an eruption that occurred shortly (~1 hour) after initial unrest.
Mount Ontake (Japan), 2014	Unanticipated (small) phreatomagmatic eruption occurred with no recognized warning, despite an extensive monitoring network (Kato et al., 2015).	No. Hindsight analysis indicates subtle anomalies that were not recognized at the time of eruption or that required time-intensive laboratory analyses. 57 individuals were killed.
Villarrica (Chile), 2015	Open-vent volcano experienced seismic and infrasonic anomalies and elevated lava lake activity for months leading up to a short-lived paroxysmal eruption. The activity represented the first explosive eruption in more than 30 years.	Yes. Alert levels were incrementally raised by Chilean authorities and tourists were kept away from the hazard zone. The severity and duration of the paroxysmal event was not anticipated.
Calbuco (Chile), 2015	Closed-vent volcano, quiet since 1972, erupted suddenly and intensely with a VEI 4 eruption.	No. Anomalous seismicity was noticed prior to eruption. In hindsight, only hours of limited seismicity preceded eruption. The monitoring network was limited.

NOTE: VEI, Volcano Explosivity Index.

used to inform probabilistic assessment (e.g., Ogburn et al., 2016).

Due to the pervasive lack of robust monitoring data and the limitations of models used to forecast eruptions, volcano monitoring agencies typically issue qualitative alerts. The USGS provides four levels of alerts: normal (or background), advisory (signs of unrest), watch (escalating unrest), and warning (dangerous eruption under way).[1] Every active or potentially active volcano in the United States is assigned a value on this scale. Such alert levels can be a starting point for civil authorities to activate plans for alert, evacuation, and shutdown of critical infrastructure, on time scales of hours to months (Aspinall et al., 2003; Punongbayan et al., 1996; Voight, 1988).

[1] See https://volcanoes.usgs.gov/vhp/about_alerts.html.

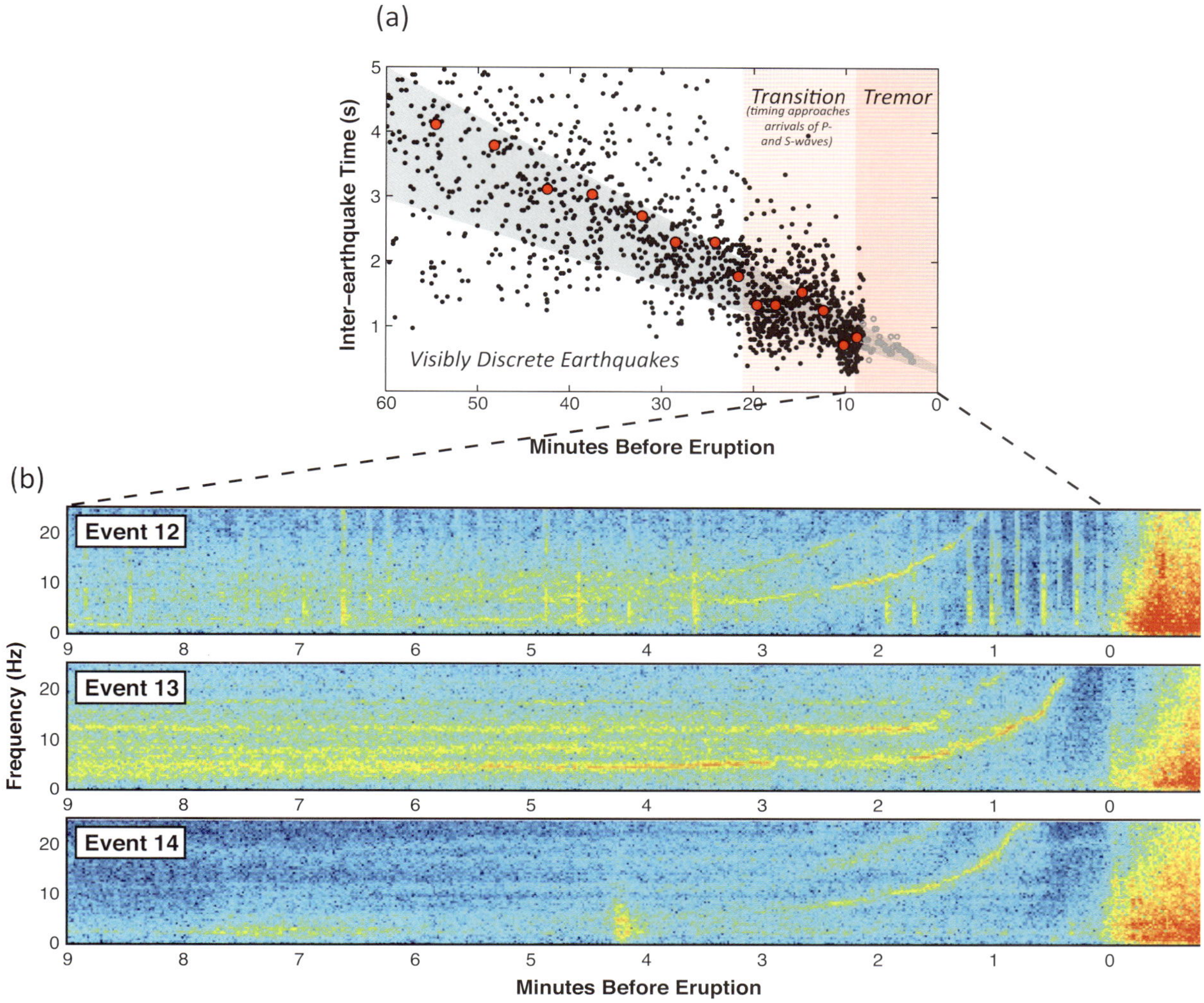

FIGURE 3.1 Changes in the rate of earthquakes and the frequency content of their seismic waves are used to forecast eruptions at well-monitored volcanoes. (a) Changes in earthquake intervals and tremor preceding explosions during the 2009 eruption of Redoubt volcano, Alaska. The interearthquake period (black dots) grades into a tremor period (gray) leading up to explosion. Red dots are temporal averages. (b) Spectrogram of gliding tremor leading up to explosions. Note the quiet period immediately prior to explosion. SOURCE: Modified from Hotovec et al. (2013).

A growing array of precursory phenomena signal unrest preceding most eruptions. Technological advances and the expansion of monitoring infrastructure at some volcanoes over the past few decades allow quick detection and interpretation of signals of unrest days to months prior to an eruption (e.g., Ewert et al., 2005). Eruptions at monitored volcanoes generally occur with at least a few hours of warning in the form of anomalous seismicity (Figure 3.1), ground deformation, hydrologic changes, and/or heat flux and gas emissions (Tables 1.1 and 1.2; Aiuppa et al., 2007). Integrated studies that combine gas emission data, ground deformation data, seismic signals, and novel petrologic techniques provide compelling evidence of magma movement months to weeks before some eruptive episodes (Kahl et al., 2013; Figure 3.2). Novel analytical techniques applied to continuous seismic signals have made it possible to detect subtle changes in seismic wave speed (Figure 3.3), interpreted as magma pressurization and ascent in the mid- to shallow crust (e.g., Brenguier et al., 2008; Obermann et al., 2013) prior to eruption, and to track dike propagation in the shallow crust without the difficulty of locating individual earthquakes during intense activity (Taisne et al., 2011). Other signals that magma is entering the shallow crust include phreatic explosions; swarms of shallow earthquakes; low-frequency earthquakes and volcanic tremor (e.g., Chouet and Matoza, 2013); changes in fluid discharge, chemistry, and temperature (e.g., White and McCausland, 2016); short-wavelength deformation; and SO_2/CO_2 gas ratios that deviate significantly from the baseline recorded during quiescence (e.g., Figure 3.2). Such signals are thought to indicate an increased likelihood of an eruption (e.g., Moran et al., 2011).

Anticipating eruptions within minutes to hours before they occur is often based on rapidly intensifying or changing indicators of unrest, such as the onset of strong volcanic tremor, ground tilt, or hydrologic changes (e.g., Figure 3.1). For example, repeating earthquakes transitioning into "gliding" volcanic tremor consistently preceded explosions of Redoubt volcano, Alaska, in 2009 and have since been linked to frictional

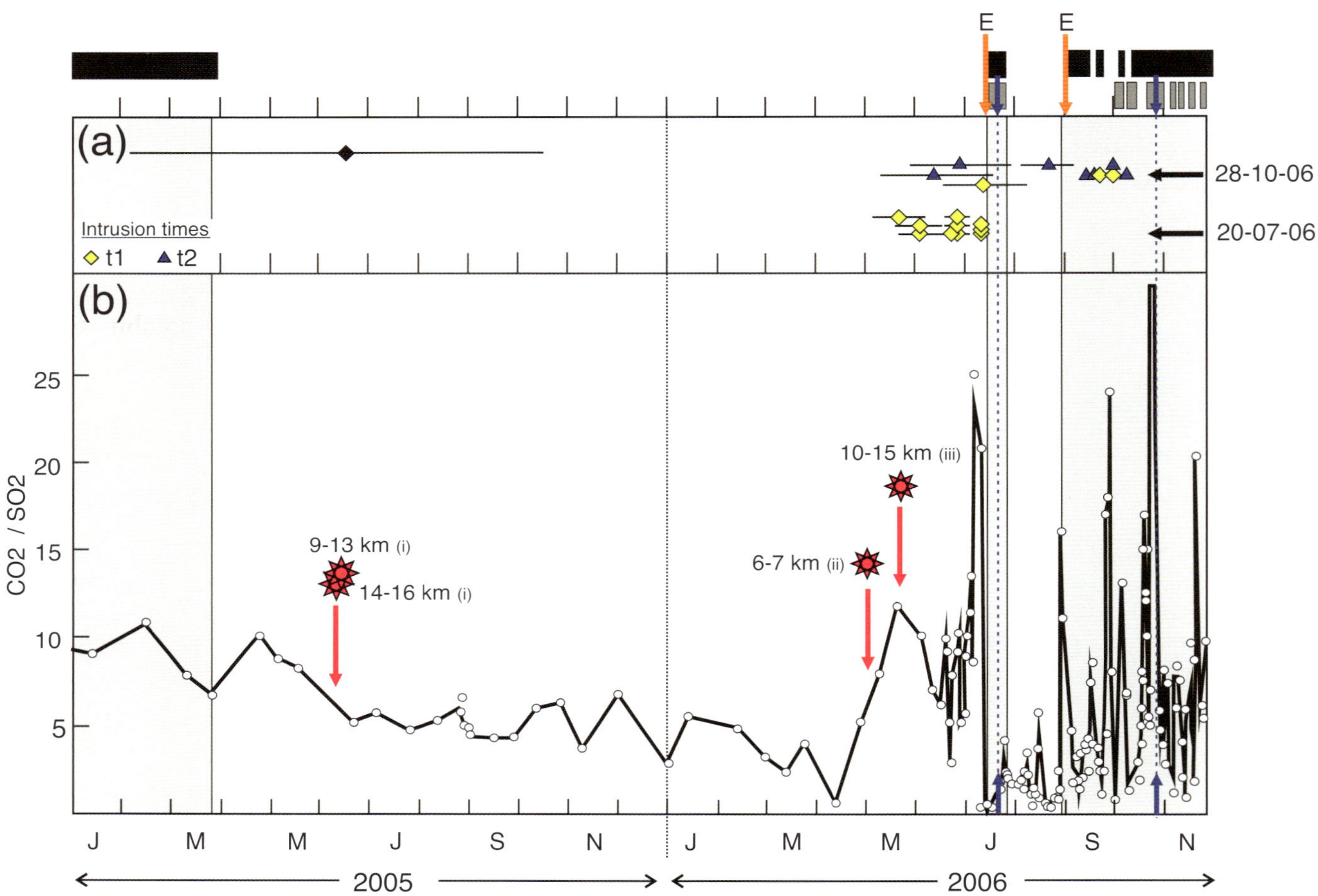

FIGURE 3.2 The ratios of chemical species in volcanic gas are sensitive indicators of magma ascent, and so can be used to forecast eruptions. For the 2006 eruption of Mount Etna, zonation of crystals collected after the eruption revealed details of the timing of several magma intrusions (a) shown by black and yellow diamonds and blue triangles, especially the onset of intrusion during April–May 2006, prior to the main eruption. The CO_2/SO_2 ratio monitored in volcanic gas emissions at the surface (b) black line with open circles) shows marked increases from a background value of about 5 to about 12 during the same time period just before the main eruption sequence, which erupted the zoned crystals shown in (a). This pre-eruption time period is also characterized by seismic events (magenta stars) occurring at 6–7 km and 10–15 km depths, likely indicating the intrusion of magma near those depths, altering the solubility of CO_2 and SO_2 in the magma and triggering the zoned growth of crystals. Orange arrows show the main eruption onset (E) and black arrows show the onset of individual eruptive events on July 20 and October 28, 2006. Vertical grey bars mark periods of eruptive activity; white areas are times of quiescence. SOURCE: Data from Kahl et al. (2013).

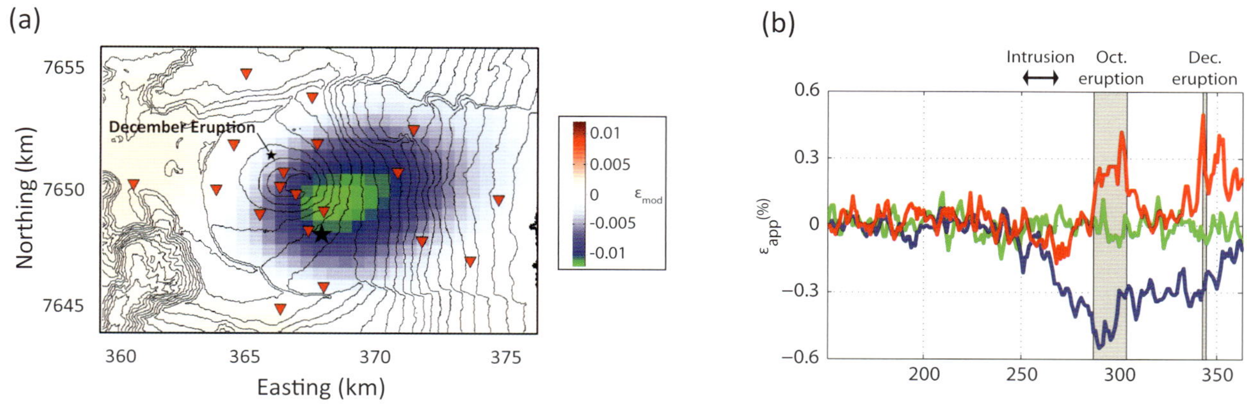

FIGURE 3.3 Ambient seismic noise is an emerging technique to detect magma in the subsurface, as well as changes in the stress state in the neighboring crust, by mapping subtle changes in seismic wave speed. (a) Map showing eruption locations and baselines between stations at Piton de la Fournaise, Reunion Island. Colors represent fractional changes in wave speed prior to the October 2010 eruption. (b) Relative change in seismic wave speed (horizontal axis, days) for the three baselines shown in (a) during 2010. Gray bands are periods of eruption. SOURCE: Modified from Obermann et al. (2013).

faulting in a highly stressed region of the conduit (Dmitrieva et al., 2013). Similarly, an exponential increase in the number of earthquakes has been interpreted as an exponential increase in the probability of eruption (Endo and Murray, 1991).

Forecasting the magnitude, style, and duration of an eruption remain major challenges for short-term forecasting. Recent efforts to analyze eruption databases for indicators of eruption volume are promising (e.g., Bebbington, 2014), with the caveat that the record of observed eruptions is biased toward small to moderate events at frequently active volcanoes, whereas the prehistoric geologic record is biased toward larger eruptions (Kiyosugi et al., 2015).

In general, detecting the onset of unrest has been far more successful than anticipating the evolution of a volcanic eruption once it has begun. Future short-term eruption forecasts must become more adept at incorporating disparate geophysical and geochemical data gathered during ongoing eruptions to create ensemble forecasts that anticipate possible changes in eruptive activity. As indicated in Table 3.1, key questions during eruptions concern the likely duration of eruptive activity, the nature of pulsatory or intermittent activity, the significance of a hiatus in eruptive activity (is the eruption over or has it only paused?), and changes in the style of activity (e.g., switch from explosive to effusive activity).

Key parameters in dynamic forecast models include the location, composition, and volatile content of the magma as well as mass fluxes of magma and gases. Once an eruption commences the combination of eruption flux and geodetic data can be used to constrain total magma chamber volume, pressure, and volatile content (e.g., Anderson and Segall, 2013; Mastin et al., 2009a). Potentially, erupted rocks and minerals could be analyzed immediately to provide information on the pressure, temperature, volatile content, and composition of the deep magmatic system feeding the eruption. Active and passive source seismic experiments with high-density coverage will continue to improve four-dimensional imaging of the volcano's plumbing system potentially in near real time (e.g., Kiser et al., 2013; Ulberg et al., 2014). Finally, remotely measured gas compositions combined with thermodynamic modeling, melt inclusion volatile contents, and solubility data could help constrain magma depth and quantity (e.g., Edmonds et al., 2001; Iacovino, 2015).

Many episodes of unrest do not culminate in eruption and better assessments of the proportion of unrest episodes that end with magma intrusion into the crust are needed (Phillipson et al., 2013). On the other hand, some explosive magmatic eruptions, such as the 2015 VEI 4 eruption of Calbuco, Chile, are preceded by surprisingly little seismicity (Romero et al., 2016). Relatively small explosive eruptions may be triggered when gas pathways are sealed by formation of a magma plug or precipitation of minerals in the hydrothermal system. This sealing process can be manifested by fluctuations in gas emissions, tilt, or long period seismicity (e.g., Cruz and Chouet, 1997; Fischer et al., 1994; Johnson et al., 2014; Nishimura et al., 2012; Rodgers et al., 2015; Stix et al., 1993; Voight et al., 1998).

Short-term forecasts are generally considered successful only when they lead to evacuation of exposed assets or populations from the hazard zone in a timely manner (Winson et al., 2014). Short-term forecasting depends greatly on the quality and quantity of ground- and space-based monitoring infrastructure (Section 1.4), the length and completeness of the baseline monitoring record, and the ability to interpret these data in a timely manner using some combination of experience as well as numerical and empirical models (Clarke et al., 2013; Peltier et al., 2005). In practice, short-term forecasting using empirical or statistical models of time series is hampered by limited sample size (for example, the limited number of times similar activity has been observed previously or the limited number of instruments deployed on a volcano). Short-term forecasts based on physics and chemistry models, whether deterministic or stochastic, are not yet used in practice due to model complexity and recalcitrant model parameters. As a result, short-term forecasts are not routine.

3.2 LONG-TERM FORECASTING

Long-term forecasts are used to estimate the likelihood and magnitude of eruptions over the life cycle of a volcano. These forecasts are relevant for land use planning over time scales of years to decades (Marzocchi and Bebbington, 2012) to more than tens of thousands of years for proposed underground nuclear waste repositories (e.g., Yucca Mountain, Nevada). Developing long-term forecasts requires reconstructing a volcano's eruptive chronology through field study (e.g., Hildreth et al., 2012) and radiometric dating. Difficulties arise due to a lack of sufficient age determinations and a variety of biases, including bias toward large events preserved in the geologic record (e.g., Kiyosugi et al., 2015), preservation bias influenced by climate, bias toward the best mapped regions of Earth, and bias toward the most recent events that are most prevalent at the surface.

Tephrachronology and deposit mapping are the most important tools for understanding magnitudes and frequencies of past eruptions and for inferring potential future activity, including large-magnitude, infrequent events (Crandell and Mullineaux, 1978; Newhall et al., 1996; Power et al., 2010; Sherrod et al., 2008). For example, annually laminated lake sediments reveal more than 100 small VEI 2 events at the basaltic and currently open-vent volcano Villarrica over the last 600 years (Van Daele et al., 2014), but field mapping indicates that a VEI 5 caldera-forming eruption occurred in the last 10,000 years—a significant and high-impact departure from the historical record. Pinatubo's 1991 eruption (Box 3.1), which was the largest of the last 100 years (Newhall et al., 1996), was anticipated based on field mapping of voluminous ignimbrite deposits of older eruptions. Thus, analysis of the geologic record (Chapter 2) and models of eruption processes to interpret the geologic record (Section 1.7) are critical to long-term forecasts (see Section 3.2).

A key problem is how to transform observations and models of the long-term behavior of the crust and mantle into long-term forecasts of magma ascent and eruption. For example, how does recognition of an electrically conductive body in the crust or mantle change long-term eruption forecasts for the next year or decade? Images of the crust and mantle developed from seismic tomography (Figure 2.4), magnetotellurics, geochemical models, and other technologies help us delineate the presence of magma in the subsurface, but the images are static and difficult to relate to the comparatively instantaneous process of dike ascent. One solution to this problem lies in modeling. That is, rather than simply recognizing a seismic tomographic anomaly in the mantle, the challenge is to create dynamically consistent models of how that anomaly changes the probability of magma ascent and eruption on a scale relevant to individual volcanoes. The problem would be relatively simple if a correlation could be identified between a single variable, say, seismic velocity perturbation in the subsurface (Figure 3.3), and eruption rate at the surface. Such direct correlations of single parameters have not yet been identified, and it is likely that future models will rely on a range of observations.

3.3 FORECASTING ERUPTION HAZARDS

Understandably, most people living near volcanoes are less concerned about whether the volcano will erupt than with the consequences of eruption. Underestimating eruption consequences has contributed to the worst volcano disasters, such as at Nevado del Ruiz, Colombia, in 1985 when lahars killed tens of thousands of people (e.g., Voight, 1990).

Most forecasts of eruption hazards depend on numerical models that simulate transport phenomena—such as the development of eruption plumes, pyroclastic flows, tephra fallout, lava flows, and lahars—given that a specific type of eruption has occurred (Section 2.3). These models can be tuned to account for a range of erupted volumes, informed by mapping. Monte Carlo simulations are used to estimate the conditional probability that a flow will inundate a specific area, or that tephra fallout will exceed a given thickness (e.g., Favalli et al., 2009; Iverson et al., 1998; Jenkins et al., 2012; Wadge et al., 1994). These conditional probabilities are also used to set priorities for instrument deployment and to help authorities formulate evacuation plans and other responses to volcanic activity.

Hazard maps are developed from a combination of geologic data and numerical models to display the forecast impacts of volcano eruptions. Maps can be based on specific scenarios or probability models (e.g., Neri et al., 2015; Figure 3.4). Currently, most hazard maps identify zones that have been inundated in the past. Because the geologic record is biased, the community is moving toward model-based hazard maps, using Monte Carlo simulations and models such as those described in Section 2.4. This approach places a high premium on model validation and verification, on how to use the geologic record to formulate model inputs, and an unbiased understanding of the life cycle of volcanoes.

For syn-eruptive forecasts, both the forecasts and the hazard maps are updated during ongoing activity. For example, the maps may update areas likely to be inundated given ongoing lava flow activity (e.g., Cappello et al., 2016). Tremendous potential exists for assimilating remotely sensed data into numerical models (Section 2.3) during eruptions to provide critical updates to hazard forecasts. Stimulated by the 2010 Eyjafjallajökull eruption that disrupted air traffic over the Atlantic and much of Europe, significant progress has been made in using satellite data and dispersion models (Section 2.3) to characterize volcanic ash emissions and mass eruption rates, and to forecast and track plume trajectories after an eruption has begun (e.g., Bursik et al., 2012; Merucci et al., 2016; Pavolonis et al., 2013; Stohl et al., 2011). A more pressing need is to use these observational methods and models to accurately forecast ash concentration in airspace downwind of the volcano in the days following an eruption. Similarly, emerging remote sensing technologies, including near-real-time four-dimensional morphological mapping and new space-borne lidars (e.g., Hughes et al., 2016), will likely improve syn-eruptive forecasts, which are crucial for identifying potential changes in eruptive activity, change in topography during eruptions, and the likely duration of eruptive events.

3.4 STEPS IN A PROBABILISTIC HAZARD ASSESSMENT

Probabilistic forecasts of volcanic eruptions are intended to account for uncertainties about when a volcano will erupt, the magnitude of the event, and the risks to people and infrastructure. Both short-term forecasts, prepared when eruption precursors are observed, and long-term forecasts, prepared before there are signs of volcanic unrest, generally follow the same steps (Aspinall et al., 2016):

1. *Develop a conceptual model of how the volcano and its magmatic system work*, using diverse geologic, geochemical, and geophysical data, drawing on patterns of activity at the volcano or analogous volcanoes. In preparing probabilistic forecasts, it is essential to focus on the types of activity that are possible, given how magma is stored and ascends in a particular system. Models address questions such as are vents distributed, or will future eruptions likely occur from a single vent? What are the likely products of volcanism? What are the likely volumes of future eruptions?

2. *Assess rates of activity*. Long-term forecasts use historical observations, radiometric dates, stratigraphy, and mapping to construct a chronology of past volcanic eruptions. A statistical model is then used to transform the chronology into a forecast of future activity. The

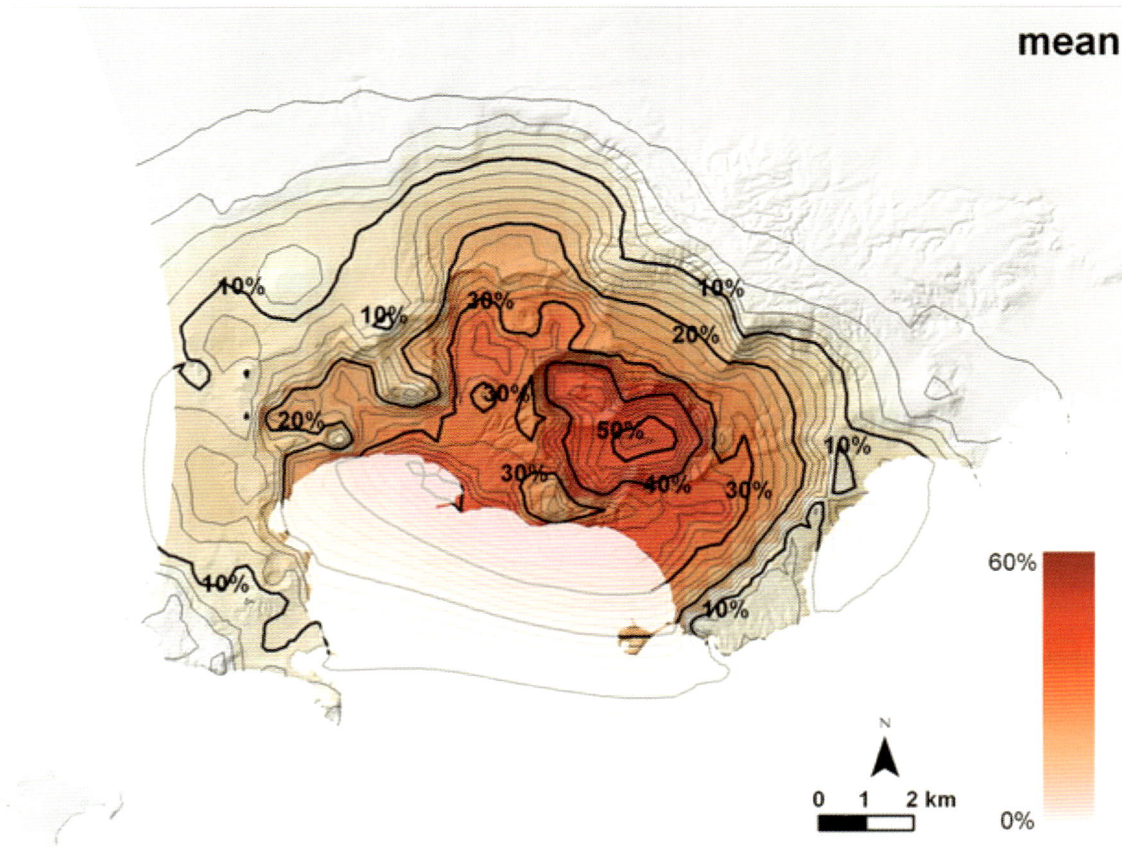

FIGURE 3.4 Long-term probabilistic forecasts can be summarized on hazard maps, showing the relative hazard to different geographic areas around a volcano. This hazard map shows the probability of pyroclastic density current inundation (contoured as percentile, and superimposed on a shaded digital elevation model) given an explosive eruption of Campi Flegrei caldera, in the densely populated region west of Naples, Italy. The map is constructed using models of expected vent location within the caldera and a model of pyroclastic density current runout, weighted by expert elicitation. Approximately 500,000 people live within the high-exposure region (>5 percent). SOURCES: Modified from Bevilacqua (2016) and Neri et al. (2015).

key sources of uncertainty are incompleteness in the geologic record and changes in eruptive behavior over time. Short-term forecasts are sensitive to changes in unrest and use information such as changes in rate of earthquakes or seismic energy release, deformation, or gas flux.

3. *Assess the potential location of activity*, particularly the locations of future vents. Long-term forecasts use the distribution of past vents, sometimes augmented by geologic or geophysical data, to create statistical models of probable vent locations. Short-term forecasts use geophysical or geochemical data to forecast potential dike intrusion and vent locations. However, even with high-resolution networks of instruments, the location of vents may remain highly uncertain.

4. *Assess the potential magnitude of activity*. For long-term forecasts, magnitude is estimated from past events. Volumes of past eruptions, for example, can be used to create a probability density function of volume. Short-term forecasts of magnitude are not the norm, although the eruption volume has sometimes been roughly estimated from the magnitudes of geophysical signals in the context of the geologic record of past eruptions (Anderson and Segall, 2013).

5. *Assess the potential impacts of activity*. Numerical models are used to estimate how far volcano products such as lava flows and tephra will extend from eruptive vents, given an eruption of a specific magnitude and style. The output of these models is usually probabilistic—for example, the likely mass loading due to

tephra accumulation at a specific location, given the volume and duration of the eruption, and other model parameters.

These steps can be summarized graphically with an event tree (Marzocchi et al., 2008; Neri et al., 2008; Woo, 2008). Nodes of the tree are defined as events (e.g., the volcano erupts, the magnitude is VEI 2, and a lava flow is produced).[2] Different nodes in a given branch are alternative events with their own probabilities, often assigned by expert judgment (Aspinall et al., 2003). Another common approach is a logic tree, which relies on the types of models discussed in this report. In logic trees, the nodes are models and alternative models for recurrence rate or vent location are each represented as a node on the graph (Figure 3.5). The transition probability is the weight assigned to each model. By assigning weights to ensemble models and calculating the probable outcomes, the sensitivity to model assumptions can be assessed directly. Event trees are easier to use and faster to implement than logic trees. Consequently, logic trees have historically been used for long-term forecasts, and event trees have been used for short-term forecasts.

3.5 FUTURE ADVANCES

Linking Monitoring and Process: Moving Toward Physics-Based Forecasting Models

Cutting-edge data analysis leading to improved understanding of how signals in monitoring data reflect key volcanic processes is critical for improving forecasting accuracy and moving beyond pattern recognition toward physics- and chemistry-based forecasting models. Particularly important are geophysical and geochemical analytical techniques that image changes in space and time, including the following:

- Documenting ambient noise and shear-wave splitting observations of wave speed changes prior to eruption by conducting experiments at more volcanoes, and correlating changes with changes in deformation and other geophysical measurements such as gravity and electrical resistivity;
- Integrating continuous Global Positioning System (GPS) and frequent interferometric synthetic aperture radar (InSAR) time series to elucidate changes in magma reservoir pressure both between and prior to eruptions;
- Testing models of volcanic source excitation by, for example, correlating seismicity with stress changes inferred from deformation observations and/or changes in gas volume or chemistry;
- Analyzing chemical and physical changes in volcano hydrothermal systems as eruption precursors and acquiring syn-eruptive measurements to evaluate eruption progress (e.g., transitions from phreatic to magmatic eruption); and
- Using continuous high-temporal-resolution and high-spatial-resolution volcanic plume gas composition and flux measurements to test models of changes in magma reservoir permeability, volatile content, redox, and temperature prior to and during eruptions.

An improved understanding of seismic wave generation, including low-frequency earthquakes and tremor, could allow these signals to be incorporated into dynamical models. Changes in stress, documented by volcano tectonic earthquakes and changes in seismic velocities, could be jointly analyzed with geodetic, gas, and gravity measurements to image subsurface magma transport (Box 3.3). Once an eruption commences, the combination of eruption flux and geodetic data can be used to constrain total magma chamber volume, pressure, and volatile content (Anderson and Segall, 2013; Mastin et al., 2009a). Eruption models conditioned on these and other observations (gas emissions, gravity, and seismicity) could be updated to yield probabilistic forecasts of future behavior (e.g., Segall, 2013), analogous to data assimilation methods in meteorology and other fields. It will be a significant challenge to develop and test such models on active volcanoes. Physical–chemical models of ash dispersal, lava flow, and, to lesser degree, pyroclastic density current inundation are more advanced and so offer more near-term promise for this approach.

Expanding Monitoring Efforts: On the Ground and from Space

Tremendous strides have been made in developing techniques to forecast eruptions in the short term.

[2] For example, see https://volcanoes.usgs.gov/vhp/forecast.html.

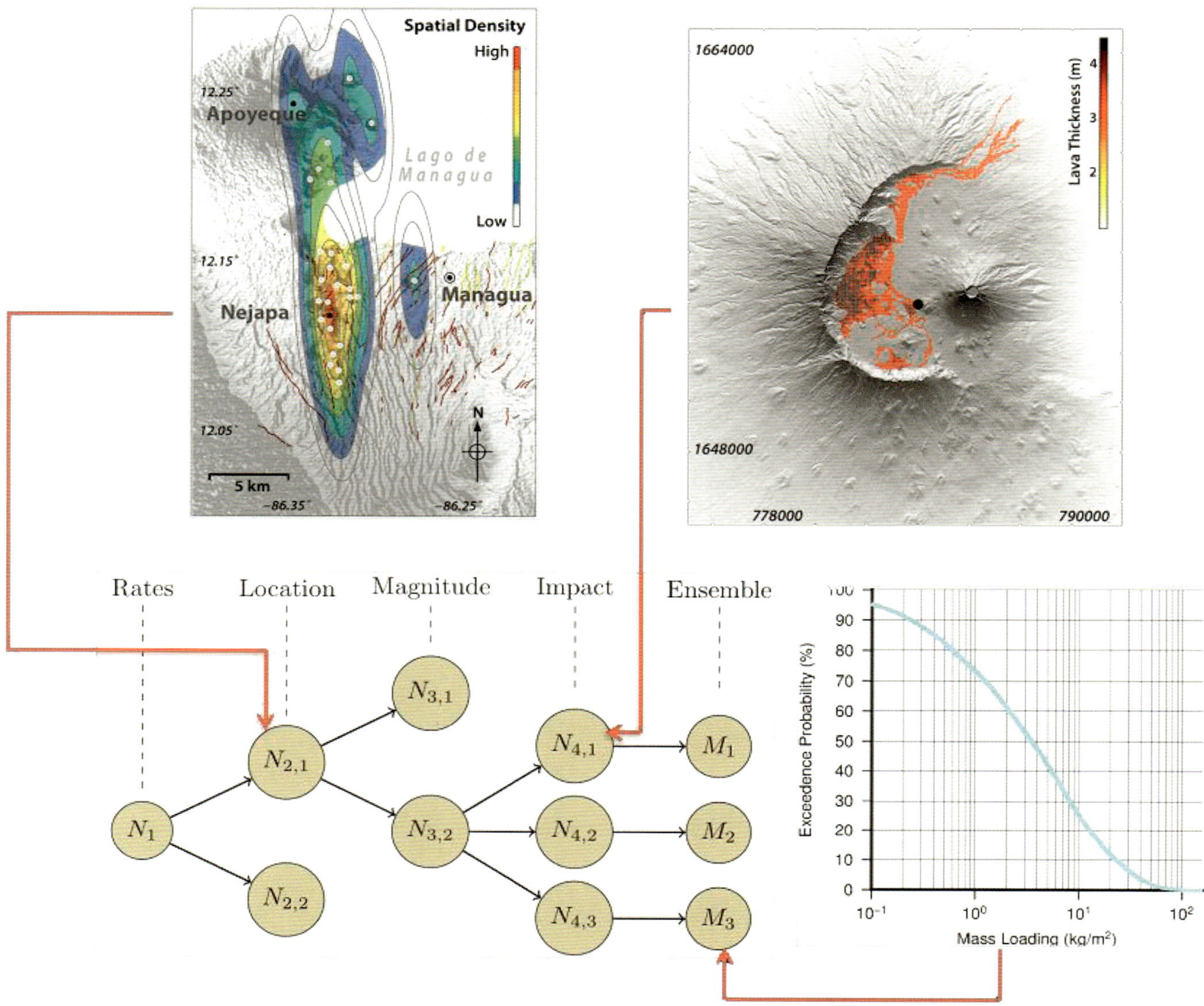

FIGURE 3.5 A pruned logic tree illustrating part of a probabilistic eruption forecast. The steps in the forecast are illustrated by nodes representing (1) a model of recurrence rate (N_1); (2) two statistical models of the location of eruptive vents ($N_{2,1}$ and $N_{2,2}$), one of which is illustrated by a probability map for potential formation of new vents (upper left); (3) two models of the magnitude ($N_{3,1}$ and $N_{3,2}$); and (4) three impact models($N_{4,1}$, $N_{4,2}$, and $N_{4,3}$), such as a numerical model of lava flow (upper right). Calculating N probabilities based on alternative models leads to weighted ensemble models, M, which are used to evaluate hazard probabilistically, often for a specific location using a hazard curve (lower right). SOURCE: Courtesy of Laura Connor, University of South Florida. Upper left and lower right figures first appeared in Sigurdsson et al. (2015).

Eruptions can be forecast using monitoring data on gas emissions, volcanic earthquakes, deformation, and other geophysical signals. Together, these phenomena are sensitive indicators of potential eruptions. Yet, in practice there is a dearth of monitored volcanoes and a paucity of coordinated monitoring studies. Even in the United States, only a subset of volcanoes are well monitored by ground-based instrumentation, and they tend to be volcanoes that erupt relatively frequently, typically producing small-magnitude events, or that are located in high-risk areas. There is a critical need for more comprehensive volcano monitoring using ground-based seismic, geodetic, and gas sensing tools. In particular, high-resolution degassing and hydrologic data are gen-

BOX 3.3
The Value of Monitoring Data

Dike injections both deform Earth's surface and induce propagating earthquake swarms. The 2014 Bárðarbunga, Iceland, dike propagated laterally 45 km over 14 days, ultimately intruding 0.5 km^3 of magma into the crust and leading to the largest eruption in Iceland in the past 230 years. Dike growth was well recorded by earthquake hypocenters, GPS data, and InSAR. The rate of dike advance varied considerably, at times propagating up to 1 km/hour, but also pausing for 80 hours (Sigmundsson et al., 2015). The seismic swarm took several sharp bends, with the trajectories apparently influenced by local topography. Time-varying GPS displacements are consistent with the dike orientation inferred from the swarm seismicity (see figure) and InSAR interferograms. Dike growth was accompanied by collapse of the Bárðarbunga caldera, as revealed by GPS, radar profiling, and radar interferometry.

Modeling of the deformation measurements indicates that most of the dike opening was shallower than 5 km, with the earthquakes concentrated slightly deeper (Sigmundsson et al., 2015). These volcano tectonic earthquakes are believed to be caused mainly by crustal stress changes, although thermal effects may also play a role, indicating an opportunity for joint interpretation of the GPS, InSAR, and seismicity data. By integrating modeling and a wide range of measurements, the Bárðarbunga eruption revealed the conditions under which caldera collapse begins, and how collapse and eruption can be coupled.

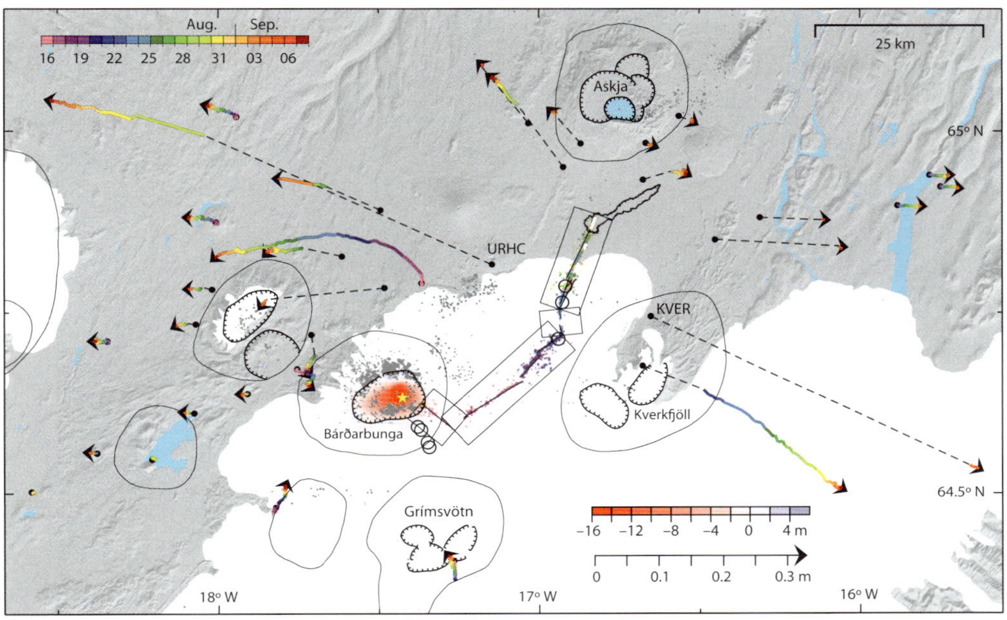

FIGURE Earthquakes between August 16 and September 6, 2014 (dots), and horizontal Global Positioning System displacements (arrows), color coded by time. Red shading indicates surface subsidence up to 16 m at Bárðarbunga caldera (which differs from the caldera subsidence caused by ice flow). Circled areas outline lava flows and eruptive fissures, inferred from a radar image on September 6, 2014. SOURCE: Modified from Sigmundsson et al. (2015).

erally less available than seismic and geodetic data, and instruments such as the miniature differential optical absorption spectrometer, multigas, and high-temporal-resolution ultraviolet and infrared cameras (Table 1.1) need to be incorporated into permanent sensor networks. When unrest begins, the basic infrastructure will need to be rapidly augmented with additional sensors and more diverse and emerging technologies, such as drones and rapid petrologic analyses. Open sharing of all data in near real time, emulating the successes of the seismologic community, will be vital.

Increased spatial and temporal resolution of satellite-borne remote sensing instruments (Table 1.2) is also crucial. Thermal sensors such as ASTER (Advanced

Spaceborne Thermal Emission and Reflection Radiometer) have high spatial resolution but low temporal resolution and so rarely provide timely observations of thermal signals such as small lava flows within craters (e.g., Reath et al., 2016). Similarly, rapid surface deformations cannot be adequately monitored with infrequent InSAR passes. For example, the planned NASA–Indian Space Research Organisation synthetic aperture radar mission provides 12-day repeat passes, which are too coarse for monitoring or documenting the evolution of eruptions. With a larger constellation of satellites, this repeat time could be reduced. It is still unclear if the increases in CO_2 emissions that can precede eruptions are detectable using current satellites (e.g., Orbiting Carbon Observatory-2) because of relatively high detection limits and low temporal resolution. Additional satellites, automated detection of anomalies via those satellites (e.g., Wright et al., 2004), as well as open access to existing data streams would significantly improve monitoring.

The paucity of well-observed large eruptions poses a different set of challenges. There is only about a 1 percent chance that a VEI ≥6 event will happen in a given year. Though relatively infrequent, the consequences of these large eruptions are grave (Figure 1.2). Thus, it is critical that the volcano science community prepare to make comprehensive and high-quality observations of the next major eruption, regardless of where on Earth it is located. It is likely that the next major eruption will occur at a completely unmonitored and poorly characterized volcano, because (1) instrumentally monitored volcanoes tend to be those which have erupted in recent history, and (2) long periods of repose may be directly correlated with erupted volume (e.g., Passarelli and Brodsky, 2012). Thus, the initial detection of precursory unrest prior to a major eruption is likely to be made via satellite or local reports of felt seismicity, ground cracking, phreatic explosions, and/or increased gas emissions, all of which may not become apparent until late in the precursory sequence. For example, precursory unrest began only a few months before the paroxysmal eruption of Mount Pinatubo in 1991. A further complication is that a large eruption may not be immediately apparent from initial precursory unrest.

Satellite-borne measurements provide a global picture of where on-land volcanoes are deforming (e.g., Fournier et al., 2010), in some cases documenting the assembly of potentially eruptible magma bodies. However, forecasting the location, timing, and magnitude of major eruptions on the basis of this information remains challenging. One way to balance the tradeoff between long repose between major eruptions and our need to mitigate their dire consequences is to work toward sparse ground-based monitoring of all potentially active volcanoes (such as one or two seismometers), noting that six instruments constitute a well-monitored volcano (Winson et al., 2014) and that monitoring strategies need to be tailored to the type of volcano in question. The utility of sparse ground-based observations can be dramatically increased by scanning for all signs of unrest, including deformation, increased heat flux, and gas emissions using satellite-borne instrumentation, ideally at least daily because of the sometimes short times between the initiation of unrest and the onset of eruption (Figure 2.5). Detection of unrest that appears to herald a major eruption would then need to be followed by rapid deployment of a dense, multiparameter network of telemetered ground-based instrumentation. Such an effort would require significant resources and advance planning, developing algorithms for automated processing and scanning of satellite data, tasking satellite-borne instruments to collect more frequent observations of restless volcanoes, a cache of ground-based instrumentation, a response plan specifying the selection of personnel and procedures for import and installation of instruments, and advance coordination with monitoring agencies worldwide.

Key Questions and Research Priorities on Forecasting Eruptions

Key Questions

- What physical parameters and processes can be identified and used to improve forecasts of whether an episode of volcanic unrest will culminate in an eruption?

- What is the best way to estimate the depth and volume of eruptible magma and use it to anticipate the magnitude of an impending eruption?
- How can precursory phenomena be used to forecast eruptive intensity and style?
- How can we forecast the duration of an eruption once it begins?
- What physical parameters of volcanic systems are most helpful in indicating which of those systems are most likely to erupt in coming decades?

Research and Observation Priorities

- Implement multidisciplinary monitoring and four-dimensional imaging of the full range of phenomena during repose, unrest, and eruption at many more volcanoes.
- Develop flexible, open-access databases of diverse observations for immediate use, and maintain them over the long term.
- Aim for seismic monitoring of each potentially active volcano and routine daily monitoring of volcanic unrest from satellites.
- Develop and test physics-based forecasting models that assimilate monitoring data and syn-eruptive observations.

4

How Do Earth Systems Interact with Eruptions?

Implicit in the goals of eruption forecasting is the assumption that improved forecasts will help to mitigate the immediate impacts of volcanic eruptions (see Chapter 3). Also critical, however, are long-term forecasts of very large eruptions and their potential for both global and long-lived impacts to Earth's environment. Volcanoes affect a host of Earth systems and vice versa. Thus, two central questions about the spatial and temporal impacts of large volcanic eruptions are (1) How do landscapes, the hydrosphere, and the atmosphere respond to volcanic eruptions? and (2) How do volcanoes respond to tectonic and climate forcing?

4.1 HOW DO LANDSCAPES, THE HYDROSPHERE, AND THE ATMOSPHERE RESPOND TO VOLCANIC ERUPTIONS?

The products of volcanic eruptions change landscapes and introduce particles and gases into the atmosphere and oceans. The immediate impacts of small to large (Volcano Explosivity Index [VEI] ≤6) volcanic eruptions on Earth systems are generally well known (Section 2.3) through observations of historical eruptions. However, the impacts of larger eruptions, such as the last super-eruption 26,000 years ago (Oruanui, New Zealand), are less well understood. Important unanswered questions are whether the impacts of very large eruptions can be anticipated by scaling up the impacts of smaller eruptions (e.g., Self, 2006) or whether the impacts of very large eruptions may be self-limiting (e.g., Oppenheimer, 2002; Timmreck, 2012; Timmreck et al., 2009). That is, will very large eruptions have unanticipated consequences for the environment and hence for human populations?

Effect on Landscapes

Volcanic eruptions can profoundly change the landscape, initially through both destructive (flank failure and caldera formation) and constructive (lava flows, domes, and pyroclastic deposits) processes, which destroy vegetation and change the physical nature of the surface (e.g., porosity, permeability, and chemistry). After explosive activity ends, secondary hazards may continue to affect local and global environments for months, years, or decades. These hazards include explosions within pyroclastic flows that occur within a few months of pyroclastic density current emplacement (Torres et al., 1996), catastrophic breakouts of lakes dammed by volcaniclastic material years after the damming event (Manville and Cronin, 2007), rainfall-generated lahars that mobilize loose pyroclastic debris for years to decades after a large eruption (Major et al., 2000; Rodolfo et al., 1996), phreatic eruptions from hydrothermal systems (e.g., Barberi et al., 1992), and sudden releases of CO_2 from volcanic lakes (e.g., Funiciello et al., 2003; Zhang, 1996).

More generally, changes in the infiltration capacity

of disturbed landscapes can greatly increase flooding and sediment transport (Pierson and Major, 2014) or, conversely, enhance remobilization of volcanic ash by wind for decades, centuries, or even millennia after a large eruption. Volcanic dust, in particular, is easily remobilized from the surface of pyroclastic deposits, as illustrated by frequent dust storms downwind of historically active volcanic regions (e.g., Liu et al., 2014; Wilson et al., 2011). Studies on the adverse effects of remobilized ash on ecosystems are few, but are increasingly recognized as an important component of ecosystem response and recovery. On even longer time scales, the landscape continues to respond by erosion and redeposition of loose surface material, rearrangement of drainage systems, regrowth of often different vegetation, and reintroduction of fauna. There are no comprehensive studies of the nature and time scales of landscape and ecosystem response, although detailed studies have traced recovery after individual volcanic eruptions (e.g., Dale et al., 2005; Del Moral and Bliss, 1993; Dull et al., 2001; Egan et al., 2016; Gunnarsson et al., 2017; Long et al., 2014; Walker et al., 2013).

Effect on the Subsurface Hydrosphere

The effects of eruptions on Earth surface processes are easy to observe and thus are fairly well quantified. Less apparent are the effects of reawakening magmatic systems on subsurface processes, particularly hydrothermal systems important for generation of energy and, over longer time spans, formation of ore deposits. Observable interactions of magmatic and groundwater systems include geophysical and geochemical signals that can be difficult to distinguish from signals of magmatic unrest. Although volcanic eruptions are commonly preceded and followed by phreatic eruptions from hydrothermal systems (e.g., Barberi et al., 1992), phreatic eruptions may also occur without warning during periods of repose and so pose a substantial forecasting challenge. Similarly, magmatic CO_2 leaked slowly into volcanic lakes can suddenly destabilize and release lethal dense gas plumes (e.g., Funiciello et al., 2003; Zhang, 1996).

Beneath the surface, magmatic–geothermal systems can generate geothermal energy and create ore deposits. Porphyry deposits in volcanic arcs provide about 75 percent of the world's copper, 50 percent of its molybdenum, 20 percent of its gold, and many metals that underpin emerging low carbon technologies (Sillitoe, 2010). It had generally been assumed that voluminous explosive volcanism is incompatible with porphyry formation. Active magmatic systems, however, are able to provide the requisite metal-bearing brines (e.g., Chelle-Michou et al., 2017), and copper ore precipitates when this brine interacts with sulfur-rich gases released from the underlying magmatic system (Blundy et al., 2015). This newly emerging understanding posits an active role for magmatism, and raises new questions about the timing of magmatism and ore formation.

Effect on the Atmosphere and Climate

Large volcanic eruptions can inject enough H_2O, CO_2, SO_2, and other volatiles (e.g., halogen species) into the upper troposphere and stratosphere to influence atmospheric chemistry and climate (Robock, 2000; Figure 4.1). Although CO_2 emitted from erupting and passively degassing volcanoes is the major pathway for mantle-derived CO_2 to enter the atmosphere (Kelemen and Manning, 2015), it is a minor component of the global mass of atmospheric CO_2 (Burton et al., 2013). For this reason, CO_2 release from all but the very largest eruptions is unlikely to change climate significantly (Self et al., 2014), although methane and CO_2 release from igneous intrusions in carbon-rich sediment can greatly increase gas emissions (e.g., Aarnes et al., 2010; Svensen et al., 2007).

The short-term effects of explosive volcanic eruptions on climate arise from the injection of volcanic SO_2 into the stratosphere where it transforms to sulfate aerosols that can persist for years, backscattering sunlight and cooling Earth's lower atmosphere and surface (Robock, 2000; see Section 2.3). Emissions of SO_2 from human activities and volcanoes, including diffuse emissions from nonerupting volcanoes, are shown in Figure 4.2. Volcano location plays an important role, with tropical eruptions being more capable of producing global impacts because seasonal variations in the Intertropical Convergence Zone facilitate transfer of aerosols between hemispheres (e.g., Kravitz and Robock, 2011; Oman et al., 2006). For this reason, even relatively small, but frequent, injections of SO_2 into the stratosphere by moderate tropical eruptions (VEI ≤ 4)

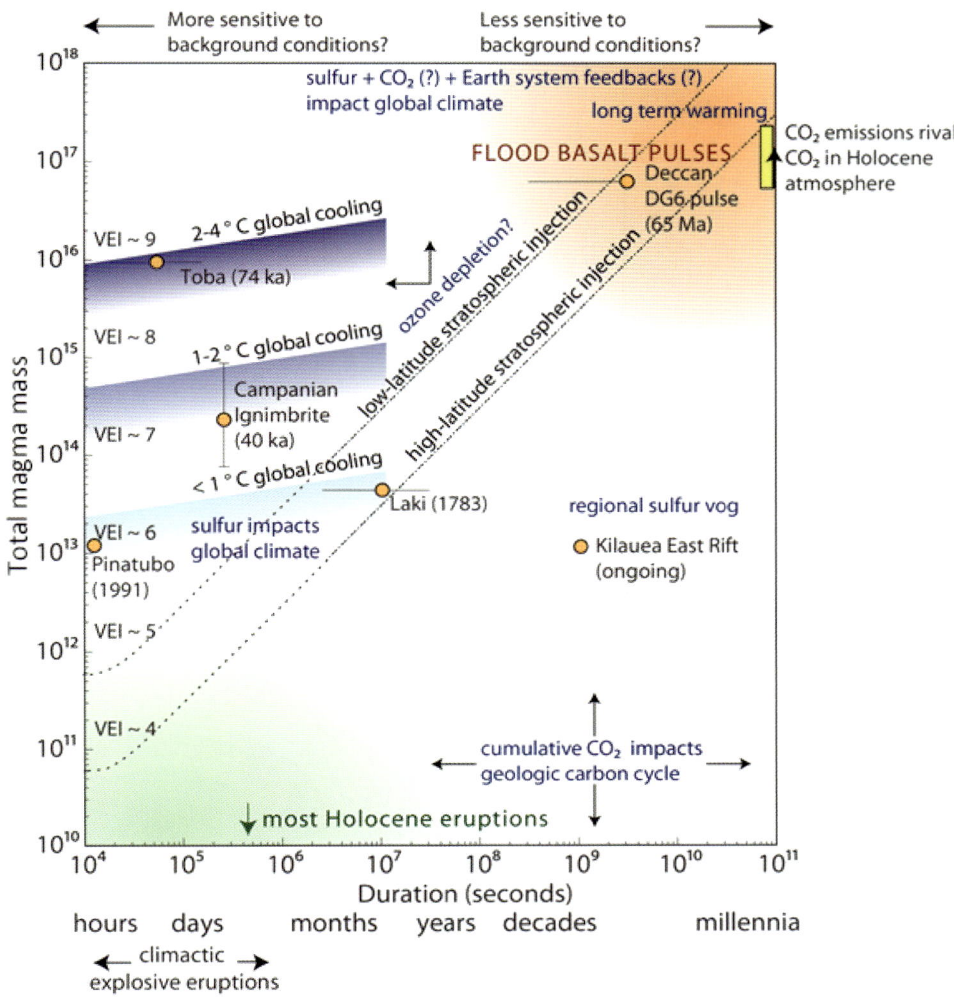

FIGURE 4.1 Volcanic eruptions of different sizes and durations have different effects on Earth's atmosphere. Words in blue identify the consequences and question marks highlight processes with the greatest uncertainty. Historical or modeled prehistoric eruptions are also shown. SOURCE: Black and Manga (2017).

may sustain the background stratospheric sulfate layer and affect climate (e.g., Santer et al., 2014; Solomon et al., 2011; Vernier et al., 2011). Less well understood are the impacts of major volcanic injections of halogen gases (Cl, Br) into the stratosphere, which could cause significant ozone depletion and generate localized ozone holes (e.g., Cadoux et al., 2015; Kutterolf et al., 2013).

The best documented global climate impact of large explosive eruptions is cooling, typically followed by winter warming of Northern Hemisphere continents, as illustrated by the 1991 eruption of Pinatubo (McCormick et al., 1995; Robock, 2000). In that event, ~10^4 teragrams of erupted magma injected 30 teragrams of aerosols into the stratosphere, the largest stratospheric loading of the past century (Figure 4.1). The negative radiative forcing caused largely by stratospheric sulfate aerosols resulted in a global tropospheric cooling of 0.2°C relative to the baseline from 1958–1991. Adjusted for the warming effect of the El Niño–Southern Oscillation (ENSO), the overall temperature decrease was 0.7°C. This temperature decrease is similar to those estimated for other sulfur-rich eruptions, such as Krakatau (1883) and Tambora (1815) in Indonesia and El Chichon (1982) in Mexico. Such temperature anomalies are short lived, so that by 1993 the tem-

FIGURE 4.2 Map of anthropogenic and volcanic SO_2 sources in East Asia and the western Pacific region based on Ozone Monitoring Instrument satellite data collected in 2005–2007. SO_2 detected over East Asia is mostly anthropogenic SO_2 emissions from China; the other SO_2 sources are mostly due to passive volcanic degassing. Significant volcanic SO_2 emissions can be seen in Japan, the Mariana Islands, the Philippines, Indonesia, Papua New Guinea, and Vanuatu. SOURCE: Based on data from Fioletov et al. (2016).

perature anomaly caused by the Pinatubo eruption had already decreased to −0.1°C (McCormick et al., 1995).

The relationship between cooling and large explosive eruptions is complex and includes not only the effect of SO_2 gas but also the effects of other emitted material (particularly H_2O, halogens, and ash), as well as the details of atmospheric chemistry that control the production and size of volcanic aerosols (e.g., LeGrande et al., 2016; Timmreck, 2012; Timmreck et al., 2009). For example, SO_2 is a greenhouse gas that could counteract the cooling effect of sulfate aerosols (Schmidt et al., 2016). Thus, the balance between SO_2 and aerosols in different parts of the atmosphere is complicated, as is the resulting climate response.

Large explosive eruptions can also affect global circulation patterns such as the North Atlantic Oscillation and ENSO (Robock, 2000), although the mechanism(s) by which this happens are not well understood (LeGrande et al., 2016). Finally, eruptions have been linked to substantial but temporary decreases in rainfall and river discharge (e.g., Oman et al., 2006; Trenberth and Dai, 2007) and the occurrence of tropical cyclones in the North Atlantic (Guevara-Murua et al., 2015). Documentation of the atmospheric impact of recent explosive eruptions provides important constraints for testing short-term climate model predictions and for exploring the effects of proposed geoengineering solutions to global warming (e.g., Robock et al., 2008, 2009).

Large effusive eruptions have a somewhat different effect on the atmosphere because of their long durations (e.g., Schmidt et al., 2016; Thordarson and Self, 2003). Basaltic eruptions, in particular, can be both voluminous and long lived, and can therefore affect local, regional, and possibly global climate. Historical examples from Iceland, such as the Laki eruption of 1783–1784 and the Bárðarbunga eruption of 2014–2015, provide an interesting contrast. The former had a regional (Northern Hemisphere) impact in the form of dry fogs of sulfuric acid (H_2SO_4), while the latter produced dangerously high local levels of SO_2. The difference reflects not only the larger volume of the Laki eruption, but also the season (summer versus winter) because sunlight plays an important role in the oxidation of SO_2 to H_2SO_4 (Gislason et al., 2015; Schmidt et al., 2010). In the extreme, the large volume and long duration of ancient flood basalts may have perturbed the atmosphere over time scales of decades to centuries to even millennia (Figure 4.1).

The effects of injecting large amounts of water by volcanic eruptions into the dry stratosphere could affect climate by accelerating the formation of sulfate aerosol by OH radicals or by decreasing the ozone formation potential of the system (Glaze et al., 1997; LeGrande et al., 2016). Studies of very large flood basalt eruptions suggest that both the formation of sulfate aerosols and the depletion of ozone played a significant role on climate over Earth's history (Black et al., 2014). These examples emphasize the need to better characterize plume gas and aerosol chemistry as well as coupling of gas-phase chemistry with aerosol microphysics in climate models. Because satellite-based remote sensing observations of volcanic gases are heavily biased toward SO_2 (e.g., Carn et al., 2016), obtaining a complete volatile inventory for explosive eruptions required for a full chemistry simulation of volcanic plumes is still a major challenge.

Effect on the Oceans

Large eruptions affect Earth's oceans in a variety of ways. Volcanic ash may be a key source of nutrients such as iron and thus capable of stimulating biogeochemical responses (Duggen et al., 2010; Langmann et al., 2010). During the week following the 2003 VEI 4 eruption of Anatahan, Northern Mariana Islands, for example, satellite-based remote sensing detected a 2–5-fold increase in biological productivity in the ocean area affected by the volcanic ash plume (Lin et al., 2011). These impacts can be particularly pronounced in low-nutrient regions of the oceans. A more indirect and longer-term impact of very large volcanic eruptions is caused by the rapid addition of CO_2 and SO_2 to the atmosphere, which affects seawater pH and carbonate saturation. Carbon-cycle model calculations (Berner and Beerling, 2007) have shown that CO_2 and SO_2 degassed from the 201-million-year-old basalt eruptions of the Central Atlantic Magmatic Province could have affected the surface ocean for 20,000–40,000 years if total degassing took place in less than 50,000–100,000 years. Ocean acidification from the increased atmospheric CO_2 may have caused near-total collapse of coral reefs (Rampino and Self, 2015). Rapid injection of large amounts of CO_2 into the atmosphere by volcanic eruptions also provides the best analog for studying the long-term effects of 20th-century CO_2 increases on ocean chemistry. Targeted investigations of these large eruptions have the potential to establish quantitative estimates of the volatile release and residence in the atmosphere as well as the effects on ocean acidification, carbon saturation, coral mortality, and biodiversity.

Over the long term, large eruptions can release thousands of gigatons of methane from organic-rich sediments. Light $\delta^{13}C$ signatures interpreted to represent such a release (Svensen et al., 2009) have been recognized in carbon isotope stratigraphic records at the Permian–Triassic (252 Ma) and Triassic–Jurassic (201 Ma) boundaries, as well as in the Paleogene (56 Ma; Saltzman and Thomas, 2012). The latter represents a well-documented thermal maximum associated with extensive volcanism that accompanied the opening of the North Atlantic Ocean. Reconstructing the volcanic carbon emission record through geologic time and assessing the potential for large releases of reduced carbon from organic sediments is challenging and requires a firm understanding of the processes that currently degas carbon and other volatiles to the atmosphere and how those signatures may be preserved in the geologic and ice core records.

Finally, some secondary volcanic hazards are generated in the ocean. Tsunamis can be generated directly by explosive submarine eruptions (e.g., Fiske et al., 1998), or indirectly by volcanic flows (pyroclastic, lahar) or debris avalanches produced by volcano flank collapses (e.g., Paris, 2015). Even small volcano-triggered tsunamis can produce significant waves (e.g., Day, 2015).

Key Questions and Research Priorities on the Response of Landscapes, the Hydrosphere, and the Atmosphere to Volcanic Eruptions

Key Questions

- How can we extrapolate observations from witnessed eruptions to anticipate the immediate and long-term effects of very large events?
- What feedbacks occur among the atmosphere, the hydrosphere, and the geosphere in the aftermath of very large eruptions?
- Under what conditions do volcanic eruptions have drastic consequences on the atmosphere and oceans?
- How do coupled magmatic and hydrothermal processes transport heat and fluids to create energy resources and ore deposits?

Research and Observation Priorities

- Increase real-time and long-term measurements of surface processes to quantify landscape evolution after eruptions.
- Monitor hydrothermal systems during periods of repose and unrest.
- Document secondary hazards, and develop models and forecasting tools for these hazards.
- Deploy satellite instruments with increased sensitivity to passive and eruptive volcanic CO_2 emissions.
- Integrate models for gas-phase chemistry and aerosol physics to account for feedbacks among volcanic, atmospheric, and ocean processes.
- Exploit high-resolution geochronology and environmental impacts preserved in ice cores and marine and lacustrine sediment to decipher eruption history, including unwitnessed very large eruptions.

4.2 HOW DO VOLCANOES RESPOND TO TECTONICS AND CHANGES IN CLIMATE?

Volcanic eruptions can be triggered when the pressure in a subsurface magma body exceeds the confining pressure in the surrounding crust, or when underpressure initiates collapse. The latter includes a contribution from surface loading (e.g., ice sheets). Active volcanoes are therefore sensitive to changes in stress, particularly those systems that are "primed" for eruption (Bebbington and Marzocchi, 2011). An external forcing mechanism that either increases magmatic overpressure or reduces the confining pressure can potentially trigger an eruption. The sources of such perturbations operate on time scales that range from near-instantaneous stress changes associated with tectonic processes such as earthquakes, to longer-term variations due to climate change such as changes in sea level and melting of ice sheets. A deeper understanding of external stimuli (tectonics, earthquakes, changes in sea level or glaciers) provides an important test of mechanisms for melt accumulation and triggering thresholds (Figure 4.3) and is necessary for improved hazard mitigation.

Tectonics

Tectonics influences volcanism by controlling the composition and amount of magma generated in the mantle and the thickness of the crust and the stresses that hinder or promote magma intrusion and ascent. Quantifying these connections would benefit from a better understanding of the properties of the crust that host magma bodies as well as the conditions that enable the propagation of dikes (Section 2.1). For example, large, silicic magma bodies that can produce caldera-

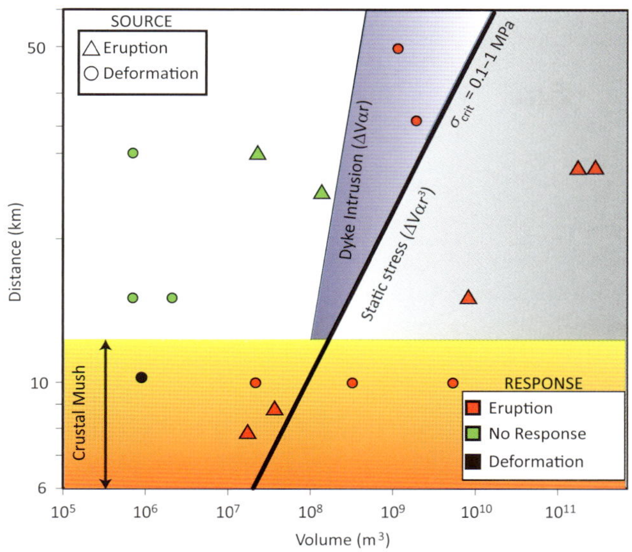

FIGURE 4.3 One volcanic eruption may trigger another at a nearby volcano, depending on volume and distance. Each point represents a volcano pair separated by the distance shown on the y-axis; the volume of magma erupted or intruded is shown on the x-axis. The pairs are represented by a "source" volcano that either showed signs of unrest (deformation) or erupted, and a "response" volcano that erupted (red), showed no response (green), or deformed (black). "No response" pairs are volcanoes that appear to have experienced triggered activity in the past (for example, Eyjafjallajökull and Katla, Iceland). Two coupling mechanisms are modeled: the blue region is defined by intrusion volumes ΔV proportional to distance r for a constant area A, as would be expected if the coupling occurred via a lateral dike (limiting value defined by $A = 10^4$ m^2); the gray region models a point source such that stress changes (σ) decay as $\Delta V/r^3$, with a limiting value for coupling delimited by a critical σ = 0.1–1 MPa (depending on crustal properties). Over short distances (~10 km), volcano–volcano interactions are probably controlled by processes that act within shared crustal mush zones (shaded orange region). SOURCE: Modified from Biggs et al. (2016).

forming eruptions are more likely to develop in thicker crust, whereas more frequent eruptions of less evolved magmas are more likely to develop in thinner, extended crust (e.g., Cembrano and Lara, 2009). There are many exceptions, however. For example, one of Earth's most frequently active silicic volcanic systems, the Taupo volcanic zone (New Zealand), is located in an extensional area. Tectonic stresses also affect magma storage and the size of eruptions (e.g., Robertson et al., 2016).

Tectonics also influences the morphology and stability of volcanoes. Volcanoes may develop on large tectonic faults (e.g., Socompa; Wadge et al., 1995) or generate faults around their base by gravitational and magmatic deformation (e.g., Etna; Acocella and Neri, 2005). Movement on tectonic faults intersecting volcanic edifices may increase the risk of flank collapse and the generation of debris avalanches, but at the same time may inhibit magmatic processes by relieving stress (e.g., Ebmeier et al., 2016). Regional stresses and faults may control the alignment of dikes, but the extent to which ambient stresses are modified by the development of magma reservoirs (e.g., Andrew and Gudmundsson, 2008; Karlstrom et al., 2009) and loading by volcanic edifices (e.g., Pinel and Jaupart, 2003) remains an open question.

Earthquakes

On a global scale, volcanism and large earthquakes are strongly spatially correlated. Most of Earth's explosive volcanoes are adjacent to subduction zones, which also generate the largest earthquakes. Temporal coincidences between earthquakes and eruptive activity have been documented since at least the writings of Pliny (his encyclopedia published in the 1st century AD). Analysis of recent earthquake and eruption catalogs shows a spike in volcanic eruptions within a few days after major (M >8) earthquakes, hinting at short-term eruption triggering at distances of many hundreds of kilometers from the epicenter (e.g., Linde and Sacks, 1998; Manga and Brodsky, 2006; Walter and Amelung, 2007). Eruption rates in the southern Andes may have increased for up to 12 months following some large earthquakes (Watt et al., 2009). However, large earthquakes do not always trigger volcanic eruptions. For example, neither the 2010 Maule nor the 2011 Tohoku earthquakes, which were of large magnitude and occurred in active and well-instrumented volcanic arcs, have been linked to triggered eruptions, perhaps because few volcanoes are "critically poised" and susceptible to triggering at any given time. The possibility of delayed triggering (e.g., the 1991 Pinatubo eruption 11 months after the M 7.8 1990 Luzon earthquake) becomes increasingly difficult to establish with time after an earthquake (Hill et al., 2002).

Persistently active volcanoes such as Merapi, Indonesia, may be particularly prone to triggered responses (e.g., Walter et al., 2007). The orientation of the earthquake focal mechanism with respect to distal volcanoes may also determine whether a triggered response occurs (e.g., Delle Donne et al., 2010). Eruptions have been attributed to earthquake-induced compression (e.g., Bonali et al., 2013; Feuillet et al., 2011; Nostro et al., 1998) or expansion of the crust (e.g., Fujita et al., 2013; La Femina et al., 2004; Walter and Amelung, 2007), nucleation or growth of bubbles (e.g., Crews and Cooper, 2014), mobilization of crystal-rich magmas by dynamic strains (e.g., Sumita and Manga, 2008), initiation of convection (e.g., Hill et al., 2002), and resonance phenomena (e.g., Namiki et al., 2016) in magma chambers. On longer time scales, earthquake-triggered ascent of deeper magmas or gases may play a role. Despite decades of study, however, the mechanisms through which seismic waves and static stress changes initiate eruptions and influence ongoing eruptions, even on short time scales, remain unknown.

Earthquakes can also trigger noneruptive unrest (seismicity, gas emissions, and changes in hydrothermal systems) at volcanoes (e.g., West et al., 2005). Indeed, hydrothermal systems are particularly sensitive to earthquakes (e.g., Ingebritsen et al., 2015). The availability of decadal or longer time series of satellite observations have facilitated investigation of links between volcanic unrest and earthquakes, especially for volcanoes without ground-based instruments. These observations reveal a range of noneruptive volcanic responses to earthquakes, including ground deformation, changes in surface heat flux, induced volcanic seismicity, and hydrologic changes (e.g., Delle Donne et al., 2010; Harris and Ripepe, 2007). Some responses suggest that eruption is less likely. Subsidence recorded at several Chilean and Japanese volcanoes following the 2010 Mw 8.8 Maule, Chile (Pritchard et al., 2013) and the 2011 Mw 9 Tohoku, Japan (Takada and Fukushima, 2013), earthquakes was attributed to coseismic release of hydrothermal fluids and enhanced subsidence of a hot, weak plutonic body, respectively. Deep long-period seismicity also decreased at Mauna Loa after the 2004 Mw 9.3 Sumatra earthquake (Okubo and Wolfe, 2008).

Volcanoes can also influence other volcanoes nearby (e.g., Linde and Sacks, 1998). Coupled eruptions have been documented, with pairs occurring within 50 km of each other (e.g., Biggs et al., 2016; Figure 4.3). The ability to predict and explain volcano responses to earthquakes and other volcanoes would be a significant advance that would aid in the interpretation of persistent unrest, such as Long Valley, California.

Climate

Although it is well understood that volcanic eruptions can impact climate (Section 4.1), relatively little attention has been paid to the potential impacts of future climate change on volcanic activity and hazards (Tuffen, 2010). On various time scales (annual to millennial), volcanoes and volcanic regions may respond to the slow surface deformation associated with seasonal and climatic cycles, such as the growth and melting of glaciers and ice sheets, and changes in sea level (e.g., Jellinek et al., 2004; Maclennan et al., 2002; Mason et al., 2004; Mather, 2015; McGuire et al., 1997; Rawson et al., 2016; Tuffen, 2010; Watt et al., 2013). Surface pressure changes induced by these processes can affect rates of decompression melting in the mantle, drive magma ascent through deformation of the crust, or lead to volatile exsolution and eruption.

Identifying correlations between volcanic activity and climate cycles relies on accurate and complete catalogs of eruptions and intrusions. Major eruptions (VEI >5) are infrequent, but their occurrence is usually, although not always, well preserved in geologic or proxy records (e.g., Rougier et al., 2016). Smaller eruptions (VEI 0–3) are more frequent and hence provide better statistics, but catalogs of such events are incomplete (e.g., Watt et al., 2013). Seasonal fluctuations of up to 50 percent of average eruption rates occur in some regions for small (VEI 0–2) eruptions (Mason et al., 2004). This fluctuation is attributed to surface deformation associated with the seasonal transfer of water between the oceans and landmasses, with volcanic eruptions more likely during periods of surface pressure change.

Large-scale melting of ice can affect the timing of eruptions. Increases in volcanic activity lag ice retreat by several thousand years at stratovolcanoes in California and Chile (Jellinek et al., 2004; Rawson et al., 2016), whereas volcanic activity in Iceland accelerated more quickly following the last deglaciation (e.g., Maclennan et al., 2002). Although glacial unloading is effectively instantaneous on geologic time scales, the lag times probably reflect the variable depth of magma supply and the transit time through the crust. At some arc volcanoes, observed lag times are shorter for eruptions of silicic magmas, which reside in shallow crustal magma chambers, than for less evolved magmas that are replenished by decompression melting in the mantle (e.g., Jellinek et al., 2004; Rawson et al., 2016).

Melting of ice leads to rising sea levels, but the volcanic response to sea-level change may promote or suppress eruptions depending on volcano type and location (McGuire et al., 1997). At mid-ocean ridges, changes in magma production may be recorded in seafloor topography (Crowley et al., 2015) and may provide CO_2-driven feedbacks with 10^5-year time lags (Burley and Katz, 2015). Hence, the feedbacks between volcanism, ice removal, and sea-level rise may be global (e.g., Huybers and Langmuir, 2017) but may also be highly variable on local and regional scales.

Changing sea level may indirectly affect eruptions by affecting flank collapse or other mass wasting events (e.g., Coussens et al., 2016). In addition, unloading the volcano may initiate eruptions (e.g., Cassidy et al., 2015). The interrelationship between flank collapse, climate, and volcanic eruptions is best deciphered from the marine sediment archive, accessible by deep sea drilling.

Although volcanic responses to glacial cycles and sea-level changes are likely the dominant climatic influence on volcanism, weather and climate can impact volcanism in other ways. Volcanic activity can be triggered by rainfall (e.g., Matthews et al., 2009; Violette et al., 2001), and there is evidence that the likelihood of volcanic flank collapse may increase in a wetter climate (e.g., Deeming et al., 2010). Future climate change may also shift the extent and/or location of the tropical rain belt, potentially decreasing eruption column heights and the ability of plumes to cross the tropopause and deliver materials to the stratosphere (e.g., Aubry et al., 2016). Our ability to forecast volcanic eruptions and their impacts in the context of a changing climate is therefore contingent on an improved understanding of the feedbacks between volcanic activity and other Earth systems.

Key Questions and Research Priorities on the Response of Volcanoes to Tectonics and Changes in Climate

Key Questions

- When and why do volcanoes interact with each other and respond to tectonics?
- How does melting ice and sea-level change affect volcanic activity?
- What are the positive and negative feedbacks between volcanism and climate change, and will they be important in the 21st century and beyond?
- How do we know when a volcano is poised for eruption?

Research and Observation Priorities

- Expand volcano monitoring to elucidate the relationship between earthquakes and hydrothermal and volcanic systems.
- Construct accurate chronologies of eruptions coupled with records of local ice and lake volume, and sea level.
- Investigate volcanic responses to climate change on time scales from glacial–interglacial cycles to extreme weather events.
- Develop improved physical models of how magma generation, storage, and eruption are affected by external influences.

5

Strengthening Volcano Science

The questions and priorities highlighted in this report are complex and multifaceted. They require perspectives on volcanism that span scales from individual crystals to entire volcanic arcs. Advances rely on instrumentation as varied as laser ablation mass spectrometers, broadband seismometers, and satellite sensors. New technology promises to provide critical insights on previously inaccessible parts of volcanoes and on eruptions.

Making new discoveries and improving understanding depend on the ability to undertake interdisciplinary research and provide interdisciplinary training. Shared infrastructure, resources, and data would accelerate the pace of progress in developing models and making critical measurements. Coordinated responses to eruptions globally would help overcome observational biases. Effective collaborations and partnerships among academia, volcano observatories, and government agencies would maximize the scientific return from monitoring data and improve eruption forecasts.

5.1 ENHANCING INTERDISCIPLINARY COLLABORATION

Volcano science is interdisciplinary. Addressing fundamental questions requires integrating diverse types of observations from geophysics, geology, geochemistry, geodynamics, and remote sensing. Efforts to improve forecasting involve research in statistics. And modern models for volcanic eruptions involve high-performance computing and collaboration with engineering science and applied mathematics. Research and understanding in each of these areas has advanced to the point that few individuals have expertise spanning more than one of these disciplines. Communication of research results is well supported through dedicated scientific journals, societies, and conferences. Supporting volcano scientists in collaborative interdisciplinary research is more challenging, however, and requires funding for cross-disciplinary research projects and for fostering sustained and substantive discussion and collaborations across disciplines.

Current core science funding programs at the National Science Foundation (NSF) are broken down into historical subdisciplines,[1] making it challenging to support multi-investigator interdisciplinary research. A few funding programs at NSF have supported interdisciplinary research involving volcanoes, including the hazards aspects of eruptions (previously the hazard elements of SEES [Science, Engineering, and Education for Sustainability], currently PREVENTS [Prediction of and Resilience against Extreme Events])

[1] Core science programs in NSF's Division of Earth Sciences are EarthScope, Geobiology and Low-Temperature Geochemistry, Geomorphology and Land Use Dynamics, Geophysics, Hydrologic Sciences, Petrology and Geochemistry, Sedimentary Geology and Paleobiology, and Tectonics. See https://www.nsf.gov/funding/programs.jsp?org=EAR.

and collaborations with mathematics (previously through Collaboration in Mathematical Geosciences [CMG]). However, these programs were not aimed at advancing our understanding of the processes that govern the storage, ascent, and eruption of magma. A successful model for interdisciplinary research is NSF's CSEDI (Cooperative Studies of the Earth's Deep Interior) program, which supports collaboration between geochemistry, geodynamics, mineral physics, geomagnetism, and seismology (similar fields to those in volcano science) to understand the evolution and dynamics of Earth's deep interior.

True collaboration between disciplines requires support for sustained exchange of ideas, challenges, and opportunities, beyond simply funding collaborative projects. Successful models for multidisciplinary problem solving often involve thematic meetings, centered around grand challenges. For examzple, the Southern California Earthquake Center is a collaboratory of geologists, seismologists, geodesists, modelers, and experimentalists who work together on specific goals defined each year, for understanding earthquake processes. GeoPRISMs (Geodynamic Processes at Rifting and Subducting Margins), an NSF multidisciplinary program to study continental margins, convenes Theoretical and Experiments Institutes that attack frontier problems. Gordon Conferences, research coordination networks, and summer institutes provide other avenues for interdisciplinary collaboration, with the added benefit of training for early career scientists. The payoffs include discoveries that would not otherwise emerge from a single perspective, new insights into complex processes, integration of data and models, and a community of scientists well versed in multiple fields and engaged in solving critical problems.

5.2 SUPPORTING COMMUNITY INFRASTRUCTURE

The volcano science community currently relies on a suite of analytical, computational, and experimental facilities. Community and multiuser facilities, in particular, provide opportunities and expertise to a broad range of users. Useful infrastructure improvements for volcano science range from analytical facilities to cyberinfrastructure, from satellites to long-lived field experiments. Infrastructure developed for complementary large Earth science projects (e.g., EarthScope and Subduction Zone Observatory) can also be leveraged.

Intrinsic properties of the magmatic systems that fuel volcanoes are measured using the tools of geochronology, geochemistry, rock physics, and petrology. Many facilities are hosted by single institutions, making access highly variable. These include geochemical and microanalytical facilities, and high-pressure, high-temperature experimental petrology and rock physics laboratories. Facilities to support geochronology, in particular, are critical for constraining the life cycles of volcanoes. Geochronology facilities in the United States are inadequate to meet the demand for all disciplines and are often inaccessible because of high costs (Harrison et al., 2015). Despite the scientific value (see Chapter 2), there is currently little U.S. emphasis on drilling to access the subsurface of volcanic systems, either for basic science studies or for deploying borehole instruments such as seismometers and strainmeters. This is in marked contrast to past community projects in the United States (e.g., Eichelberger, 1997; Eichelberger et al., 1984; Keller et al., 1979; Zablocki et al., 1974) and current international efforts (e.g., Bonaccorso et al., 2016; Elders et al., 2014; Sakuma et al., 2008).

Experimental facilities to study the dynamics of volcanic phenomena such as pyroclastic density currents, lava flows, and plumes are hosted by individual researchers. The large scale of many experimental models, however, could benefit from development and support of community user facilities (Valentine et al., 2011). Such large-scale experiments can provide a test bed for exploring new physical processes, validating codes, and testing new instrumentation, including in situ monitoring of flows.

Department of Energy-supported synchrotron beamlines have fueled rapid advances in spectroscopic and single-crystal measurements and microtomographic studies of volcanic materials. New and exciting applications include four-dimensional imaging of multiphase magma transport processes (e.g., Baker et al., 2012).

Access to high-performance computing, such as XSEDE (Extreme Science and Engineering Discovery Environment), NCAR (National Center for Atmospheric Research), and NERSC (National Energy Research Scientific Computing Center) facilities, are essential for state-of-the-art models. Communal cyberinfrastructure supports comprehensive databases, model development, benchmarking, and implementation (e.g., Marzocchi et al., 2008; Sparks et al., 2012). VHub[2] currently serves as a clearinghouse for such models. Community facilities, such as NSF-funded CIG (Computational Infrastructure for Geodynamics), could help improve the accessibility and user-friendliness of computational approaches by enhancing code efficiency and offering resources not available to individuals.

Finally, geophysical, geochemical, and geodetic data underpin research on active volcanoes. Key avenues for data collection include satellites that can be used for targeted observations of restless volcanoes, airborne instruments (including drones), and instrument pools for volcano-specific monitoring equipment. The NSF-funded PASSCAL (Portable Array Seismic Studies of the Continental Lithosphere) center provides seismometers for targeted campaigns, and UNAVCO provides engineering support and Global Positioning System (GPS) units for campaign measurements and permanent installation. A broader range of instruments and enhanced community coordination would maximize rapid response capabilities and permit innovative multisensor experiments on individual volcanoes.

5.3 PREPARING FUTURE VOLCANO SCIENTISTS

Improving our ability to understand and forecast volcanic behavior requires a workforce capable of communicating and integrating information across the different fields represented within volcano science. The next generation of volcano scientists must not only develop core expertise but also acquire sufficient knowledge to incorporate results from, and communicate with, volcano scientists in other disciplines. Of these skills, only the first is a common goal of traditional graduate programs. Although most graduate programs also encourage some breadth in scientific understanding, the extent varies widely. Skills for communicating across disciplines are only indirectly a part of the training of most scientists. Improved training therefore requires new ways to expand the communication proficiency of scientists, and to foster opportunities and mechanisms for developing interdisciplinary research skills while still maintaining disciplinary rigor. Similarly, quantitative skills are increasingly important, requiring training in computation and statistics.

Training interdisciplinary volcano scientists poses several challenges. First is the breadth of volcano science—few institutions can cover all aspects internally. An exception is the U.S. Geological Survey (USGS), which is only peripherally involved in graduate student training. Second is the sheer size of the United States, which means that research institutions specializing in volcano science are spread across the country (particularly Hawaii and Alaska). The physical separation means that casual exchanges and interactions among volcano scientists are less than optimal. More formal interactions take place at specialized meetings, but they are difficult to maintain over the long term. Finally, the funding structure tends to be conservative, typically encouraging discipline-specific projects (Section 5.1) and thus indirectly discouraging exploratory work between disciplines.

A variety of programs in Europe address these challenges in volcano science by enabling joint training in different disciplines and across different institutions.[3] These programs come with their own sets of challenges: They are expensive and time consuming for all participants, and they can be impractical for those with limited geographic mobility. Training networks, however, have the advantage of training a cohort of PhD students who, during their studies, will have developed what will hopefully be a lifelong network. An alternative approach is to develop a postdoctoral program that requires training in a field that is outside of, but complementary to, the PhD specialty.

On shorter time scales are focused summer schools, such as the Geophysical Fluid Dynamics Program[4] (run by the Woods Hole Oceanographic Institute, now past its 50th year) and the CIDER (Cooperative Institute for Dynamic Earth Research)[5] summer program (funded by NSF, now past its 10th year). Also important are training schools, internships, and volunteer programs that provide students the experience of working in an observatory environment. Examples of these

[2] See https://vhub.org.

[3] Some examples include the European Research Commission's Innovative Training Networks (e.g., VERTIGO, NEMOH) and the European Science Foundation (e.g., MEMOVOLC).

[4] See http://www.whoi.edu/gfd.

[5] See https://www.deep-earth.org.

programs include the international training course run by the Center for the Study of Active Volcanoes and the volunteer program run by the Hawaiian Volcano Observatory, both of which provide hands-on experience for students interested in Hawaiian volcanism.

5.4 DEVELOPING THE NEXT GENERATION OF INSTRUMENTATION AND BROADENING APPLICATIONS OF INSTRUMENTATION TO VOLCANO SCIENCE

Volcano science is grounded in a rich history of empirical observation, and new ground-based, airborne, and satellite technology are allowing many kinds of observations to be acquired more rapidly and in more detail than ever before. Substantial advances in our understanding of internal and surface processes have been achieved through acquisition and interpretation of seismic, magnetotelluric, deformation, gas, hydrologic, and thermal data, from space and on the ground (see Chapters 2 and 3). Evolving technology is permitting more interplay between measurements. For instance, seismic study has been expanded to seismoacoustics, incorporating infrasound recorded by low-frequency microphones (e.g., Arrowsmith et al., 2010). Radar is providing new insights into processes in explosive eruptions. Seismic and deformation monitoring now document a continuum of Earth motions. Thermal cameras, previously used to quantify stationary heat flow, now provide high-temporal-resolution imagery to measure both vent velocities and lava effusion rates (e.g., Patrick et al., 2014). A new generation of multispectral imaging cameras are able to measure gas concentrations in volcanic plumes at high enough spatial and temporal resolution to enable direct comparisons to seismic signals (e.g., Nadeau et al., 2011). Accompanying the benefits of expanding data sets and instrument capabilities are challenges posed by data that are heterogeneous in both space and time, making comparisons between precursory signals difficult.

Future advances in volcano science will be facilitated by sensor improvement and the deployment of global multiparameter sensor networks to capture the full range of temporal and spatial variability of volcanic activity. Experiments at model volcanoes such as Stromboli, Italy, or analog volcanoes such as geysers (Hurwitz and Manga, 2017), offer opportunities to interpret a host of geophysical signals in terms of processes and physical properties. Further developments in spectroscopy will enable higher-precision measurements of the isotopic and chemical composition of volcanic gases remotely, in situ, and in near real time. Advances in technology will make instruments smaller, cheaper, and more robust. Higher precision and spatial resolution of laboratory-based beam analytical techniques will provide finer temporal resolution of processes recorded by crystals and melt inclusions.

Several planned or proposed satellite missions would benefit the volcano science community (Davis et al., 2016). Distinguishing volcanic CO_2 from anthropogenic emissions remains challenging with current sensors, but deployment of satellite-based technology with greater vertical sensitivity to CO_2 (e.g., active laser instruments such as NASA's ASCENDS [Active Sensing of CO_2 Emissions over Nights, Days, & Seasons]) could lead to more timely detection of eruption precursors. This would be particularly effective in combination with more frequent (daily) repeat interferometric synthetic aperture radar (InSAR) measurements of ground deformation from a constellation of satellites. Regular acquisition of global, high-quality digital elevation models such as the TanDEM-X (TerraSAR-X add-on for Digital Elevation Measurement) WorldDEM[6] would facilitate studies of dynamic volcano topography and permit more accurate simulations of volcanic mass flows and improved hazard mapping.

The scale of advances in data acquisition are illustrated by changing capabilities of volcano seismology (Table 5.1); however, all of the technologies summarized in Tables 1.1 and 1.2 have undergone a similar evolution or could be developed further with adequate resources. Improved database capabilities (Section 5.5) and software are needed to complement sensor improvements and the increased volume and quality of data. Efficient archiving and extraction of time series, spectral, and image data are crucial to improve data visualization and discovery. Finally, drone technology promises to revolutionize the capabilities for data and sample collection by allowing access to inaccessible or dangerous areas or by offering previously unanticipated perspectives.

[6] See http://www.intelligence-airbusds.com/worlddem.

TABLE 5.1 Advances in Volcano Seismology

Year	Data Collection Capability
1980	Seismic data at volcanoes are analog and short period (>1 Hz). Recordings are displayed on paper helicorder plots.
1990	All seismic data are recorded digitally, but not continuously. Digital tape archiving is costly and inefficient.
2000	Seismic data are recorded continuously. Broadband seismology is revolutionizing volcano earthquake study.
2010	Digital networks have largely replaced analog systems. Large-N deployments allow development of high-resolution volcano tomographic images.
2020	What's next? Thousands of very broadband geodetic seismometers rapidly deployed by drones and delivering data in real time.

Data from many of the instruments used to monitor and study volcanoes provide insights into other hazards, such as earthquakes, landslides, and forest fires. They also yield information on subsurface processes, such as the evolution and structure of the crust, the development of geothermal systems, and the formation of ore deposits. There are thus opportunities to leverage instrumentation and networks to address a range of resource, hazard, and science questions.

5.5 SUPPORTING ACCESS TO DATA AND DATA PRODUCTS

Open data access has revolutionized some disciplines in Earth science. For example, the easy availability of waveform data has allowed for new and innovative analysis in seismology. This in turn has led to the discovery of phenomena such as tectonic tremor, as well as insights on the structure of Earth's deep interior. Similar arguments can be made for readily accessible continuous GPS data.

Databases are playing an increasingly important role in volcano science. A summary of existing volcano databases is given in Appendix A. Flexible databases allow comparisons of parameters and phenomena across many volcanoes and eruptions. For example, it would be useful to know how often phreatic explosions are followed by magmatic eruptions as well as the distribution of time intervals between these events. Data access in volcano science is inherently more challenging than for seismology, because the field involves such disparate data types, including the following:

- Historical information on past eruptions, including Volcanic Explosivity Index, eruption rate, erupted products and volume, duration of eruption, events during eruptions (e.g., explosions, pyroclastic density currents, and lahars), and stratigraphic and field relations, including deposit thickness and extent.
- Data on potential eruption precursors, such as
  - o seismicity (earthquake locations, magnitudes, moment tensors, and frequency content; tremor amplitude; real-time seismic-amplitude measurement);
  - o deformation, including "snapshots" (InSAR interferograms) and time series (GPS, leveling, tilt, strain, and sets of InSAR images);
  - o gas and fluid measurements (ground based and remotely sensed);
  - o thermal measurements (ground based and remotely sensed); and
  - o infrasound.
- Rock samples, including composition, phase assemblages, textures, and melt inclusion volatiles, ideally tied to specific eruptions.
- Physical volcanology parameters, such as deposit thickness, mass, density, and grain size and shape.
- Imaging data, including data from cameras (e.g., photos, video, and time lapse images) and geophysical techniques (e.g., seismic tomography images).
- Potential field measurements, including gravity and magnetotellurics.

Once an eruption has begun, a variety of data is generated through ground-based and remote sensing techniques, including eruption column heights and ash and gas distribution; pyroclastic density current and lava flow paths and volumes; petrology, geochemistry, and fluxes of erupting products and emitted gases; deformation; and seismicity. Accurate and frequent, possibly real-time or near-real-time, ingestion into databases would allow scientists at observatories to improve eruption modeling and forecasts of how the eruption will proceed and when it may end.

Despite the many existing databases, key information is not currently included, such as compositional information for eruptions over time; field data (e.g., maps and videos); ash fall, pyroclastic density current, lava flow, and lahar inundation maps; textural and sieve data; and samples including location information.

There are many challenges to moving toward a comprehensive volcano science database, including establishing standards for relevant data and metadata, linking databases through web services, and making a long-term commitment to maintenance. Significant efforts were required to develop standards for seismic networks, and these apply mainly to large networks such as the Global Seismic Network. Data from volcanoes are likely to be much more heterogeneous. Moreover, much of the relevant data are collected by volcano observatory staff, who have little time to disseminate them during heightened activity or may be concerned about public misunderstanding and alarm. The USGS will be making all published data publicly available, although mechanisms to share data with volcano scientists in real time have not yet been developed.

University researchers, who put enormous effort into data collection, may also be reluctant to make data freely available before publishing their interpretations. Different disciplines have different standards for data sharing. For example, there is a 2-year moratorium before seismic data must be made publicly available through the IRIS (Incorporated Research Institutions for Seismology) Data Management Center. Digital Object Identifiers provide one mechanism for acknowledging researchers' contribution to data collections.

Additional challenges are specific to the USGS Volcano Disaster Assistance Program (VDAP). VDAP is deployed outside the United States following a formal request through the U.S. Agency for International Development. USGS employees are guests of the host countries and observatories. Part of their mission is to build capacity. The pattern has been for data to remain in the host country. There is thus an inherent tension between promoting the professional careers of local scientists by giving them primacy in publishing data, and ensuring that unique data are made available to the broader scientific community and archived in perpetuity.

5.6 MAXIMIZING THE VALUE OF COLLABORATIONS BETWEEN OBSERVATORY AND ACADEMIC VOLCANO SCIENTISTS

Observatory and academic volcano scientists are currently well positioned to foster partnerships to take full advantage of rich data sets collected through monitoring and ensure that scientific gains from future major eruptions are maximized. Bringing together the different knowledge, expertise, and perspectives of these two groups will contribute to building a strong volcano science community capable of making new discoveries, developing and testing new instrumentation and monitoring techniques, and implementing more accurate and sophisticated forecasting models. USGS–academic partnerships can support the mission of the USGS by expanding the community of scientists studying volcanoes, and by training the next generation of professionals engaged in volcano science. The National Earthquake Hazards Reduction Program program[7] has been successful in promoting such partnerships for earthquake science.

Volcano observatories excel at long-term volcano monitoring, as exemplified by the recent centennial of the USGS Hawaiian Volcano Observatory (Babb et al., 2011). These activities result in long-duration data sets that could not be collected by academic researchers through standard (e.g., NSF) funding mechanisms. While the primary use of these data is to assess the current state of a volcano and its potential for eruptive activity, collaborative retrospective analyses of monitoring data have led to many new insights on the fundamental processes that govern volcanic behavior (e.g., Carey et al., 2015). Such collaborative efforts lead to scientific advances while supporting the mission of the observatories as new instruments, analytical techniques, and models become incorporated into monitoring efforts.

Observatory–academic partnerships are an ideal vehicle for training the next generation of volcano scientists through graduate student internships and postdoc positions, taking advantage of the observatories' proximity to active volcanoes, experienced and multidisciplinary staff, and exposure to the challenges of maintaining monitoring networks. Some of the fundamental questions highlighted in this report could

[7] See http://www.nehrp.gov.

be tackled through research projects cosupervised by observatory and academic partners and supported through national funding agencies. Educational exchanges would benefit both groups. Moreover, observatory personnel from developing countries could attend graduate school in the United States, ideally leading to knowledge transfer, while access to a student workforce could help observatories fully mine data archives resulting from monitoring activities.

A major challenge in understanding volcanoes is that significant leaps in understanding volcanic processes tend to occur during and immediately following rare well-observed major or otherwise significant eruptions (e.g., Mount St. Helens in 1980, Pinatubo in 1991, and Eyjafjallajökull in 2010). It is thus critical that academic and observatory partners prepare in advance to maximize comprehensive and high-quality observations of the next major eruption (IAVCEI Task Group on Crisis Protocols, 2016). Careful long-term planning during "peacetime" is key for managing complementary objectives during the crisis: forecasting and hazard mitigation, and high-quality data and sample collection. Academic partners would need to provide observatory partners with a description of capabilities and/or resources they could contribute during a crisis or periods of volcanic unrest as well as a summary of data they need to advance scientific understanding. Furthermore, best practices for the collection of critical and ephemeral data and samples need to be established in advance to avoid loss. Finally, academic scientists would need to formulate an action plan to coordinate personnel and equipment and to disseminate the resulting data and samples within the community.

5.7 BUILDING AN EFFECTIVE VOLCANO SCIENCE COMMUNITY

An effective volcano science community requires several elements, including the following:

- Support for interdisciplinary collaboration and training, which is essential to making discoveries and integrating models and measurements;
- Shared community infrastructure, which is necessary for state-of-the-art modeling, analytical facilities, monitoring, and field experiments;
- Databases that preserve and facilitate open exchange of information and hence enable exploration of the life cycle of volcanoes and improve forecasting;
- New technology and instruments that permit new detection, measurements, and sampling, including previously inaccessible parts of ongoing eruptions;
- A coordinated response by the research community to eruptions globally to overcome observational bias; and
- Volcano observatory–academic partnerships, which will accelerate the translation of basic science to applications and monitoring.

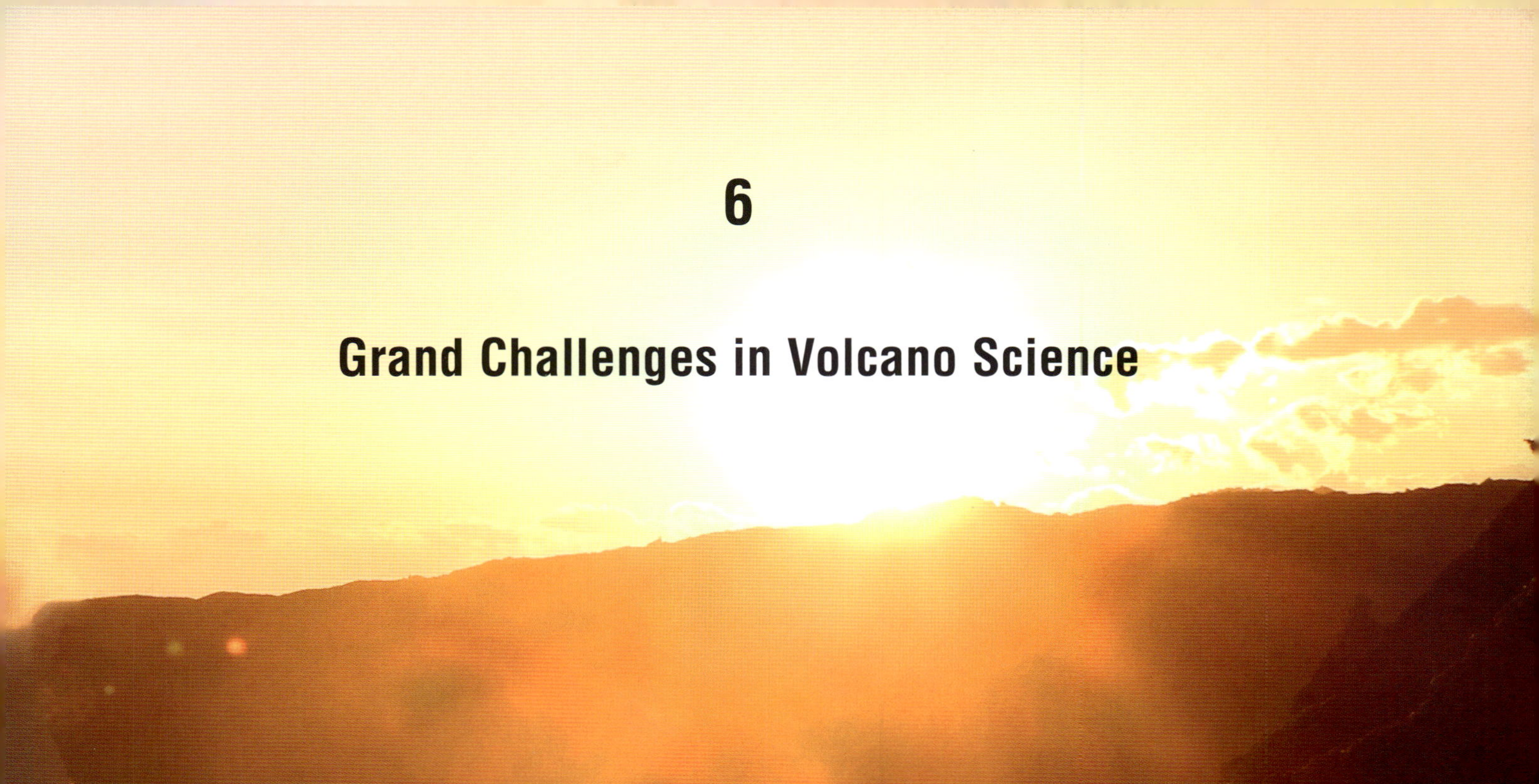

6

Grand Challenges in Volcano Science

Our understanding of the life cycle of volcanoes is poised for major advances. The field of volcano science has evolved from one dominated by a description of deposits firmly rooted in geologic traditions, to a multidisciplinary field that also exploits the latest satellite and ground-based measurements, high-performance computing, and new field and laboratory instrumentation. The key questions, research priorities, and new approaches highlighted throughout this report can be summarized by three grand challenges. These challenges are grand because they are large in scope and will have important results, and they are challenges because great effort will be needed.

> ***1. Forecast the onset, size, duration, and hazard of eruptions by integrating observations with quantitative models of magma dynamics.***

Developing conceptual models of volcanic systems as well as physics- and chemistry-based models that can inform forecasting requires the integration of data and methodologies from multiple disciplines. These include remote sensing, geophysics, geochemistry, atmospheric science, mathematical modeling, and statistics. Addressing this grand challenge also requires new understanding of basic processes, rates, and thresholds (see Chapter 2), which will come from using new instruments and approaches for exploring volcanic systems and from interdisciplinary research. National Science Foundation–supported programs that have successfully enabled cross-discipline collaboration include SEES (Science, Engineering, and Education for Sustainability), CMG (Collaboration in Mathematical Geosciences), and CSEDI (Cooperative Studies of the Earth's Deep Interior), but such programs have been underutilized in volcano science.

Understanding of eruption processes and hazards have benefited from advances in technology and computation (Section 5.4). Forecasting is critically dependent on the quality and accessibility of databases (Section 5.5). Access to and support of analytical, computational, and experimental facilities (Section 5.2) are essential for volcano science.

> ***2. Quantify the life cycles of volcanoes globally and overcome our current biased understanding.***

Determining the life cycle of volcanoes is key for interpreting precursors and unrest (see Chapter 3); revealing the processes that govern the initiation, magnitude, and longevity of eruptions (Sections 2.2 and 2.3); and understanding how magmatic systems evolve during the quiescence between eruptions (Section 2.1). However, our understanding of the volcano life cycle is spatially biased by the small number of volcanoes studied in detail, and temporally biased because large eruptions are rare in the modern instrumental era. Data from satellites and expanded ground-based monitoring networks can overcome some of these observational biases, as can extending observations to the ocean basins. A useful goal is to have at least one seismometer per volcano, complemented by extensive ground-based monitoring at a smaller number of high-priority volcanoes, global and daily satellite imaging of deformation, and the ability to measure passive CO_2 degassing from space. Geologic studies, augmented by cored scientific drilling and geophysical imaging of volcanic systems, remain necessary to understand volcanism over longer periods of time. These are large-scale projects.

Emerging technologies, including inexpensive sensors and drones and new microanalytical geochemical methods, provide previously unimagined opportunities. Monitoring strategies can be informed by the emerging understanding of volcanic processes, and can be tailored to the geological setting and expected behavior. Maintaining and expanding monitoring capabilities, and supporting the infrastructure to make historical and monitoring data available (Section 5.5), are essential for advancing understanding of volcanic processes and assessing volcanic hazards.

3. Develop a coordinated volcano science community to maximize scientific returns from any volcanic event.

Volcano science often advances substantially following well-studied eruptions. However, many eruptions occur at poorly monitored volcanoes in both populated and remote regions. The research community needs to be prepared to monitor or respond to volcanoes globally. Such preparations involve strengthening multidisciplinary research, domestic and international partnerships, and training networks (Section 5.1).

Individual academic departments in the United States are too small to support all the areas of research that must be integrated to study volcanoes. Large-scale, multi-institutional training networks and research partnerships, between government agencies and universities around the world, are critical for providing the breadth and depth of expertise needed to prepare and sustain professionals in volcano science (Section 5.3).

U.S. Geological Survey (USGS)–academic partnerships can support the mission of the USGS by expanding the available community of scientists studying volcanoes, and in training the next generation of professionals engaged in volcano science (Section 5.6). A variety of models for such partnerships exist. For example, the National Earthquake Hazards Reduction Program and the Southern California Earthquake Center have been successful in promoting partnerships for earthquake science.

In summary, huge strides have been made to understand volcanic systems on a variety of scales. It is undeniable that conceptual models of volcanic phenomena are vastly improved compared to those of a few decades ago. Yet the volcano science community is not yet adequately prepared for the next large eruption. The fundamental challenges summarized in this report will require sustained effort over years to decades, but must be addressed before eruption forecasting is routine and precise. The ongoing eruption at Bogoslof volcano, Alaska (Box 6.1) highlights these three grand challenges, why they remain timely and why they are important. The community is poised to move forward with a broad, interdisciplinary effort to obtain key data, assimilate data and models, and understand the four-dimensional structure of magmatic systems. By addressing these three grand challenges, volcano science can help quantify the global effect of eruptions and mitigate hazards, benefiting the millions of people living in volcanically active areas.

BOX 6.1
The 2016–2017 Eruption of Bogoslof Volcano, Aleutian Islands, United States

Bogoslof, a remote, mostly submarine volcano in the Aleutian Island arc began erupting in late December 2016 and activity continues as of February 2017. The Bogoslof eruption highlights several of the challenges facing volcano science. Over one month, the volcano produced numerous explosions with plumes rising 20,000–35,000 feet, posing a significant hazard to North Pacific aviation. There are no ground-based instruments (e.g., seismometers) on the volcano, and so the USGS Alaska Volcano Observatory (AVO) has been relying on distant seismometers, satellite data, infrasound, and lightning detection to monitor activity (Challenge 3). Bogoslof's submerged vent obscures any pre-eruptive thermal or gas signals, and infrasound and lightning are detectable only after eruptions have begun (Challenge 1). AVO has been able to provide early warning for only some of these hazardous events. The eruption also highlights our limited understanding of magma–water interactions and raises important questions regarding the controls on phreatomagmatic explosivity, column altitude, ash removal, and pauses (Challenge 2). In more than 20 discrete events, the emerging volcano has reshaped its coastlines repeatedly, providing snapshots of volcano–landscape interactions. The figure below shows the first evidence for an ash-rich (brown-grey) plume, almost one month into the eruptive activity.

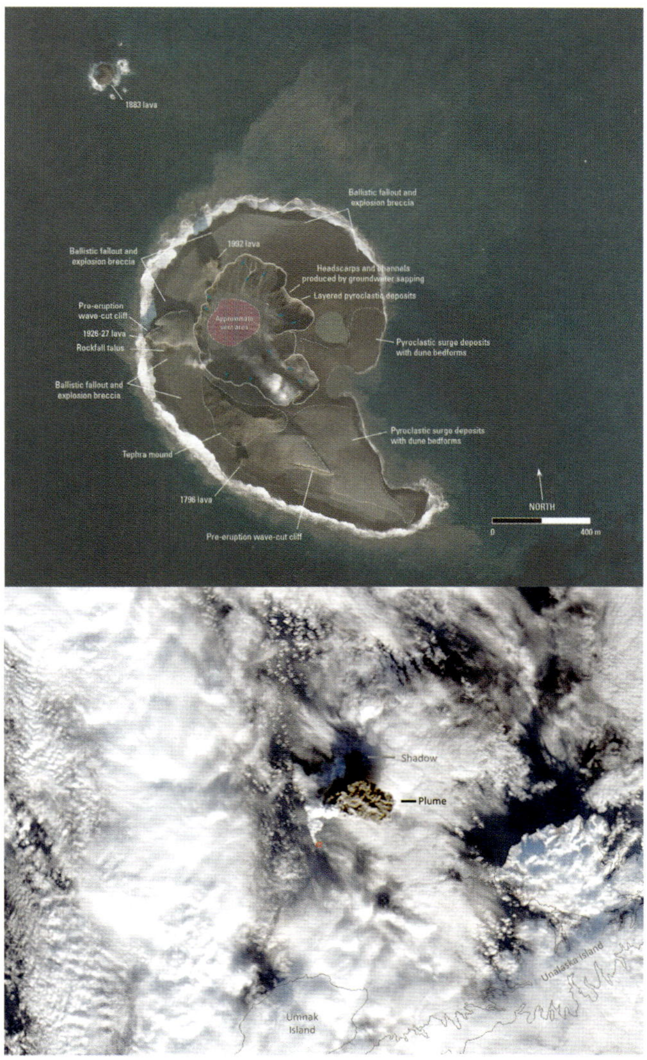

FIGURE (*Top*) Annotated aerial photo of Bogoslof volcano on January 10, 2017, showing morphological changes associated with the 2016–2017 eruption. SOURCE: USGS/AVO; www.avo.alaska.edu. (*Bottom*) NASA MODIS satellite image showing the first notably ash-rich eruption plume from Bogoslof on January 18, 2017. Note the ash-rich plume top and the white, water-rich cloud base. Ash was first sampled only about a month after eruptions began. SOURCE: NASA Visible Earth; http://visibleearth.nasa.gov/view.php?id=89476.

References

Aarnes, I., H. Svensen, J.A. Connolly, and Y.Y. Podladchikov. 2010. How contact metamorphism can trigger global climate changes: Modeling gas generation around igneous sills in sedimentary basins. Geochimica et Cosmochimica Acta 74(24):7179-7195; doi:10.1016/j.gca.2010.09.011.

Acocella, V. 2014. Great challenges in volcanology: How does the volcano factory work? Frontiers in Earth Science 2:4; doi:10.3389/feart.2014.00004.

Acocella, V., and M. Neri. 2005. Structural features of an active strike-slip fault on the sliding flank of Mt. Etna (Italy). Journal of Structural Geology 27(2):343-355; doi:10.1016/j.jsg.2004.07.006.

Aiuppa, A., R. Moretti, F. Cinzia, G. Giudice, S. Gurrieri, M. Liuzzo, P. Papale, H. Shinohara, and M. Valenza. 2007. Forecasting Etna eruptions by real-time observation of volcanic gas composition. Geology 35(12):1115-1118; doi:10.1130/G24149A.1.

Aiuppa, A., H. Shinohara, G. Tamburello, G. Guidice, M. Liuzzo, and R. Moretti. 2011. Hydrogen in the gas plume of an open vent volcano, Mount Etna, Italy. Journal of Geophysical Research: Solid Earth 116(B 10):204; doi:10.1029/2011JB008461.

Albert, H., F. Costa, and J. Martí. 2016. Years to weeks of seismic unrest and magmatic intrusions precede monogenetic eruptions. Geology 44(3):211-214; doi:10.1130/g37239.1.

Alfano, F., C. Bonadonna, P. Delmelle, and L. Costantini. 2011. Insights on tephra settling velocity from morphological observations. Journal of Volcanology and Geothermal Research 208(3-4):86-98; doi:10.1016/j.jvolgeores.2011.09.013.

Alidibirov, M., and D.B. Dingwell. 1996. Magma fragmentation by rapid decompression. Nature 380(6570):146-148; doi:10.1038/380146a0.

Allan, A.S.R., D.J. Morgan, C.J.N. Wilson, and M.A. Millet. 2013. From mush to eruption in centuries: Assembly of the super-sized Oruanui magma body. Contributions to Mineralogy and Petrology 166(1):143-164; doi:10.1007/s00410-013-0869-2.

Anderson, K., and P. Segall. 2013. Bayesian inversion of data from effusive volcanic eruptions using physics-based models: Application to Mount St. Helens 2004-2008. Journal of Geophysical Research: Solid Earth 118(5):2017-2037; doi:10.1002/jgrb.50169.

Anderson, K.R., M.P. Poland, J.H. Johnson, and A. Miklius. 2015. Episodic deflation–inflation events at Kılauea volcano and implications for the shallow magma system. Pp. 229-250 in Hawaiian Volcanoes: From Source to Surface, R. Carey, V. Cayol, M. Poland, and D. Weiss, eds. Hoboken, NJ: Wiley. doi:10.1002/9781118872079.ch11.

Andrew, R.E., and A. Gudmundsson. 2008. Volcanoes as elastic inclusions: Their effects on the propagation of dykes, volcanic fissures, and volcanic zones in Iceland. Journal of Volcanology and Geothermal Research 177(4):1045-1054; doi:10.1016/j.jvolgeores.2008.07.025.

Andrews, B.J. 2014. Dispersal and air entrainment in unconfined dilute pyroclastic density currents. Bulletin of Volcanology 76(9):852-852; doi:10.1007/s00445-014-0852-4.

Andrews, B.J., and M. Manga. 2012. Experimental study of turbulence, sedimentation, and coignimbrite mass partitioning in dilute pyroclastic density currents. Journal of Volcanology and Geothermal Research 225-226:30-44; doi:10.1016/j.jvolgeores.2012.02.011.

Annen, C., J.D. Blundy, and R.S.J. Sparks. 2006. The genesis of intermediate and silicic magmas in deep crustal hot zones. Journal of Petrology 47(3):505-539; doi:10.1093/petrology/egi084.

Aoki, Y., P. Segall, T. Kato, P. Cervelli, and S. Shimada. 1999. Imaging magma transport during the 1997 seismic swarm off the Izu Peninsula, Japan. Science 286(5441):927-930; doi:10.1126/science.286.5441.927.

Arnulf, A.F., A.J. Harding, G.M. Kent, S.M. Carbotte, J.P. Canales, and M.R. Nedimović. 2014. Anatomy of an active submarine volcano. Geology 42(8):655-658; doi:10.1130/G35629.1.

Arrowsmith, S.J., J.B. Johnson, D.P. Drob, and M.A.H. Hedlin. 2010. The seismo-acoustic wavefield: A new paradigm in studying geophysical phenomena. Reviews of Geophysics 48(4):RG4003; doi:10.1029/2010RG000335.

Aspinall, W.P., G. Woo, B. Voight, and P.J. Baxter. 2003. Evidence-based volcanology: Application to eruption crises. Journal of Volcanology and Geothermal Research 128(1-3):273-285; doi:10.1016/S0377-0273(03)00260-9.

Aspinall, W.P., S. Charbonnier, C.B. Connor, L.J.C. Connor, A. Costa, L.M. Courtland, H. Delgado Granados, A. Godoy, K. Hibino, B.E. Hill, J.C. Komorowski. S. McNutt, K. Meliksetian, S. Morita, S. Nakada, C. Newhall, S.K. Samaddar, I.P. Savov, S. Self, Y. Uchiyama, T. Wilson, T. Yamamoto, and K. Watanabe. 2016. Volcanic Hazard Assessments for Nuclear Installations: Methods and Examples in Site Evaluation. IAEA-TECDOC-1795. Vienna: International Atomic Energy Agency. Available at http://www-pub.iaea.org/MTCD/Publications/PDF/TE1795web.pdf. Accessed March 1, 2017.

Aubry, T.J., M. Jellinek, W. Degruyter, C. Bonadonna, V. Radic, M. Clyne, and A. Quainoo. 2016. Impact of global warming on the rise of volcanic plumes and implications for future volcanic aerosol forcing. Journal of Geophysical Research: Atmospheres 121(22):13326-13351; doi:10.1002/2016JD025405.

Babb, J.L., J.P. Kauahikaua, and R.I. Tilling. 2011. The Story of the Hawaiian Volcano Observatory—A Remarkable First 100 Years of Tracking Eruptions and Earthquakes. General Information Product 135. U.S. Geological Survey. Available at http://pubs.usgs.gov/gip/135. Accessed December 8, 2016.

Bachmann, O., and G.W. Bergantz. 2003. Rejuvenation of the Fish Canyon magma body: A window into the evolution of large-volume silicic magma systems. Geology 31(9):789-792; doi:10.1130/G19764.1.

Bachmann, O., and G.W. Bergantz. 2006. Gas percolation in upper-crustal magma bodies as a mechanism for upward heat advection and rejuvenation of near-solidus magma. Journal of Volcanology and Geothermal Research 149(1-2):85-102; doi:10.1016/j.jvolgeores.2005.06.002.

Baines, P.G., and R.S.J. Sparks. 2005. Dynamics of giant volcanic ash clouds from supervolcanic eruptions. Geophysical Research Letters 32(24); doi:10.1029/2005GL024597.

Baker, D.R., F. Brun, C. O'Shaughnessy, L. Mancini, J.L. Fife, and M. Rivers. 2012. A four dimensional X-ray tomographic microscopy study of bubble growth in basaltic foam. Nature Communications 31:1135; doi:10.1038/ncomms2134.

Barberi, F., A. Bertagnini, P. Landi, and C. Principe. 1992. A review on phreatic eruptions and their precursors. Journal of Volcanology and Geothermal Research 52(4):231-246; doi:10.1016/0377-0273(92)90046-G.

Barboni, M., P. Boehnke, A.K. Schmitt, T.M. Harrison, P. Shane, A.S. Bouvier, and L. Baumgartner. 2016. Warm storage for arc magmas. Proceedings of the National Academy of Sciences of the United States of America 113(49):13959-13964; doi:10.1073/pnas.1616129113.

Barnes, C.R., M.M.R. Best, B.D. Bornhold, S.K. Juniper, B. Pirenne, and P. Phibbs. 2007. The NEPTUNE Project—A cabled ocean observatory in the NE Pacific: Overview, challenges and scientific objectives for the installation and operation of Stage I in Canadian waters. Pp. 308-313 in Symposium on Underwater Technology and Workshop on Scientific Use of Submarine Cables and Related Technologies, April 17-20, 2007. Institute of Electrical and Electronics Engineers. doi:10.1109/UT.2007.370809.

Bebbington, M.S. 2014. Long-term forecasting of volcanic explosivity. Geophysical Journal International 197(3):1500-1515; doi:10.1093/gji/ggu078.

Bebbington, M.S., and W. Marzocchi. 2011. Stochastic models for earthquake triggering of volcanic eruptions. Journal of Geophysical Research: Solid Earth 116:B05204; doi:10.1029/2010JB008114.

Beckett, F.M., C.S. Witham, M.C. Hort, J.A. Stevenson, C. Bonadonna, and S.C. Millington. 2015. Sensitivity of dispersion model forecasts of volcanic ash clouds to the physical characteristics of the particles. Journal of Geophysical Research: Atmospheres 120(22):11636-11652; doi:10.1002/2015JD023609.

Behnke, S.A., R.J. Thomas, S.R. McNutt, D.J. Schneider, P.R. Krehbiel, W. Rison, and H.E. Edens. 2013. Observations of volcanic lightning during the 2009 eruption of Redoubt Volcano. Journal of Volcanology and Geothermal Research 259:214-234; doi:10.1016/j.jvolgeores.2011.12.010.

Belousov, A., B. Voight, and M. Belousova. 2007. Directed blasts and blast-generated pyroclastic density currents: A comparison of the Bezymianny 1956, Mount St Helens 1980, and Soufrière Hills, Montserrat 1997 eruptions and deposits. Bulletin of Volcanology 69(7):701-740; doi:10.1007/s00445-006-0109-y.

Benage, M., J. Dufek, and P.A. Mothes. 2016. Quantifying entrainment in pyroclastic density currents from the Tungurahua eruption, Ecuador: Integrating field proxies with numerical simulations. Geophysical Research Letters 43(13):6932-6941; doi:10.1002/2016GL069527.

Bergantz, G.W., J.M. Schleicher, and A. Burgisser. 2015. Open-system dynamics and mixing in magma mushes. Nature Geoscience 8(10):793-796; doi:10.1038/ngeo2534.

Berner, R.A., and D.J. Beerling. 2007. Volcanic degassing necessary to produce a $CaCO_3$ undersaturated ocean at the Triassic-Jurassic boundary. Palaeogeography, Palaeoclimatology, Palaeoecology 244(1-4):368-373; doi:10.1016/j.palaeo.2006.06.039.

Bertani, R. 2015. Geothermal power generation in the world 2010-2014 update report. Proceedings World Geothermal Congress, April 19-25, 2015, Melbourne, Australia. Available at https://pangea.stanford.edu/ERE/db/WGC/papers/WGC/2015/01001.pdf. Accessed December 8, 2016.

Bevilacqua, A. 2016. Doubly Stochastic Models for Volcanic Hazard Assessment at Campi Flegrei Caldera. Pisa: Scuola Normale Superiore; doi:10.1007/978-88-7642-577-6.

Biggs, J., S.K. Ebmeier, W.P. Aspinall, Z. Lu, M.E. Pritchard, R.S.J. Sparks, and T.A. Mather. 2014. Global link between deformation and volcanic eruption quantified by satellite imagery. Nature Communications 5:3471; doi:10.1038/ncomms4471.

Biggs, J., E. Robertson, and K. Cashman. 2016. The lateral extent of volcanic interactions during unrest and eruption. Nature Geoscience 9:308-311; doi:10.1038/ngeo2658.

Black, B.A., and M. Manga. 2017. Volatiles and the tempo of flood basalt magmatism. Earth and Planetary Science Letters 458:130-140; doi:10.1016/j.epsl.2016.09.035.

Black, B.A., J.F. Lamarque, C.A. Shields, L.T. Elkins-Tanton, and J.T. Kiehl. 2014. Acid rain and ozone depletion from pulsed Siberian Traps magmatism. Geology 42(1):67-70; doi:10.1130/G34875.1.

Blake, S., and B.C. Bruno. 2000. Modelling the emplacement of compound lava flows. Earth and Planetary Science Letters 184(1):181-197; doi:10.1016/S0012-821X(00)00278-8.

Blundy, J.D., K.V. Cashman, and K. Berlo. 2008. Evolving magma storage conditions beneath Mount St. Helens inferred from chemical variations in melt inclusions from the 1980-1986 and current (2004-2006) eruptions. Pp. 755-790 in A Volcano Rekindled: The Renewed Eruption of Mount St. Helens, 2004-2006, D.R. Sherrod, W.E. Scott, and P.H. Stauffer, eds. U.S. Geological Survey Professional Paper 1750. Available at https://pubs.usgs.gov/pp/1750/chapters/pp2008-1750_chapter33.pdf. Accessed December 9, 2016.

Blundy, J., J. Mavrogenes, B. Tattitch, S. Sparks, and A. Gilmer. 2015. Generation of porphyry copper deposits by gas-brine reaction in volcanic arcs. Nature Geoscience 8(3):235-240; doi:10.1038/ngeo2351.

Bonaccorso, A., A. Linde, G. Currenti, S. Sacks, and A. Sicali. 2016. The borehole dilatometers network of Mt. Etna: A powerful tool to detect and infer volcano dynamics. Journal of Geophysical Research: Solid Earth 121(6):4655-4669; doi:10.1002/2016JB012914.

Bonadonna, C., R. Cioni, A. Costa, T. Druitt, J. Phillips, L. Pioli, D. Andronico, A. Harris, S. Scollo, O. Bachmann, G. Bagheri, S. Biass, F. Brogi, K. Cashman, L. Dominguez, T. Dürig, O. Galland, G. Giordano, M. Gudmundsson, M. Hort, A. Höskuldsson, B. Houghton, J.C. Komorowski, U. Küppers, G. Lacanna, J.L. Pennec, G. Macedonio, M. Manga, I. Manzella, M. de' Michieli Vitturi, A. Neri, M. Pistolesi, M. Polacci, M. Ripepe, E. Rossi, B. Scheu, R. Sulpizio, B. Tripoli, S. Valade, G. Valentine, C. Vidal, and N. Wallenstein. 2016. MeMoVolc report on classification and dynamics of volcanic explosive eruptions. Bulletin of Volcanology 78:84; doi:10.1007/s00445-016-1071-y.

Bonali, F.L., A. Tibaldi, C. Corazzato, D.R. Tormey, and L.E. Lara. 2013. Quantifying the effect of large earthquakes in promoting eruptions due to stress changes on magma pathway: The Chile case. Tectonophysics 583:54-67; doi:10.1016/j.tecto.2012.10.025.

Brand, B.D., and A.B. Clarke. 2009. The architecture, eruptive history, and evolution of the Table Rock Complex, Oregon: From a Surtseyan to an energetic maar eruption. Journal of Volcanology and Geothermal Research 180(2):203-224; doi:10.1016/j.jvolgeores.2008.10.011.

Brand, B.D., C. Mackaman-Lofland, N.M. Pollock, S. Bendaña, B. Dawson, and P. Wichgers. 2014. Dynamics of pyroclastic density currents: Conditions that promote substrate erosion and self-channelization—Mount St Helens, Washington (USA). Journal of Volcanology and Geothermal Research 276:189-214; doi:10.1016/j.jvolgeores.2014.01.007.

Branney, M.J., and B.P. Kokelaar. 2002. Pyroclastic Density Currents and the Sedimentation of Ignimbrites. Memoir No. 27. Geological Society of London. Available at http://www.doganaydal.com/nesneler/kutuphanekitaplar/pyroclastic_density_currents_and_the_sedimentation_of_ignimbrites_.pdf. Accessed December 9, 2016.

Breard, E.C.P., G. Lube, S.J. Cronin, R. Fitzgerald, B. Kennedy, B. Scheu, C. Montanaro, J.D.L. White, M. Tost, J.N. Procter, and A. Moebis. 2014. Using the spatial distribution and lithology of ballistic blocks to interpret eruption sequence and dynamics: August 6, 2012 Upper Te Maari eruption, New Zealand. Journal of Volcanology and Geothermal Research 286:373-386; doi:10.1016/j.jvolgeores.2014.03.006.

Breard, E., G. Lube, J. Jones, J. Dufek, S. Cronin, G. Valentine, and A. Moebis. 2016. Coupling of turbulent and non-turbulent flow regimes within pyroclastic density currents. Nature Geoscience 9(10):767-771; doi:10.1038/ngeo2794.

Brenguier, F., N.M. Shapiro, M. Campillo, V. Ferrazzini, Z. Duputel, O. Coutant, and A. Nercessian. 2008. Towards forecasting volcanic eruptions using seismic noise. Nature Geoscience 1(2):126-130; doi:10.1038/ngeo104.

Brenguier, F., D. Rivet, A. Obermann, N. Nakata, P. Boué, T. Lecocq, M. Campillo, and N. Shapiro. 2016. 4-D noise-based seismology at volcanoes: Ongoing efforts and perspectives. Journal of Volcanology and Geothermal Research 321:182-195; doi:10.1016/j.jvolgeores.2016.04.036.

Brown, R., C. Bonadonna, and A. Durant. 2012. A review of volcanic ash aggregation. Physics and Chemistry of the Earth, Parts A/B/C 45:65-78; doi:10.1016/j.pce.2011.11.001.

Buck, W.R., P. Einarsson, and B. Brandsdóttir. 2006. Tectonic stress and magma chamber size as controls on dike propagation: Constraints from the 1975–1984 Krafla rifting episode. Journal of Geophysical Research: Solid Earth 111 (B12); doi:10.1029/2005JB003879.

Buisson, C., and O. Merle. 2002. Experiments on internal strain in lava dome cross sections. Bulletin of Volcanology 64(6):363-371; doi:10.1007/s00445-002-0213-6.

Burgisser, A., and W. Degruyter. 2015. Magma ascent and degassing at shallow levels. Pp. 225-236 in Encyclopedia of Volcanoes, 2nd Ed., H. Sigurdsson, B. Houghton, S.R. McNutt, H. Rymer, and J. Stix, eds. San Diego, CA: Academic Press. doi:10.1016/B978-0-12-385938-9.00011-0.

Burgisser, A., G.W. Bergantz, and R.E. Breidenthal. 2005. Addressing complexity in laboratory experiments: The scaling of dilute multiphase flows in magmatic systems. Journal of Volcanology and Geothermal Research 141(3-4):245-265; doi:10.1016/j.jvolgeores.2004.11.001.

Burgisser, A., C. Oppenheimer, M. Alletti, P. Kyle, B. Scaillet, and M.R. Carroll. 2012. Backward tracking of gas chemistry measurements at Erebus volcano. Geochemistry, Geophysics, Geosystems 13(11):Q11010; doi:10.1029/2012GC004243.

Burley, J.M.A., and R. Katz. 2015. Variations in mid-ocean ridge CO_2 emissions driven by glacial cycles. Earth and Planetary Science Letters 426:246-258; doi:10.1016/j.epsl.2015.06.031.

Bursik, M.I., and A.W. Woods. 1996. The dynamics and thermodynamics of large ash flows. Bulletin of Volcanology 58(2):175-193; doi:10.1007/s004450050134.

Bursik, M.I., M.D. Jones, S. Carn, K. Dean, A.K. Patra, M. Pavolonis, E.B. Pitman, T. Singh, P. Singla, P. Webley, H. Bjornsson, and M. Ripepe. 2012. Estimation and propagation of volcanic source parameter uncertainty in an ash transport and dispersal model: Application to the Eyjafjallajokull plume of 14-16 April 2010. Bulletin of Volcanology 74(10):2321-2338; doi:10.1007/s00445-012-0665-2.

Burton, M.R., G.M. Sawyer, and D. Granieri. 2013. Deep carbon emissions from volcanoes. Reviews in Mineralogy and Geochemistry 75(1):323-354; doi:10.2138/rmg.2013.75.11.

Büttner, R., P. Dellino, and B. Zimanowski. 1999. Identifying magma–water interaction from the surface features of ash particles. Nature 401(6754):688-690; doi:10.1038/44364.

Cadoux, A., B. Scaillet, S. Bekki, C. Oppenheimer, and T.H. Druitt. 2015. Stratospheric ozone destruction by the Bronze-Age Minoan eruption (Santorini Volcano, Greece). Scientific Reports 5:12243; doi:10.1038/srep12243.

Calder, E.S., R.S.J. Sparks, and M.C. Gardeweg. 2000. Erosion, transport and segregation of pumice and lithic clasts in pyroclastic flows inferred from ignimbrite at Lascar Volcano, Chile. Journal of Volcanology and Geothermal Research 104(1-4):201-235; doi:10.1016/S0377-0273(00)00207-9.

Calder, E.S., R. Luckett, R.S.J. Sparks, and B. Voight. 2002. Mechanisms of lava dome instability and generation of rockfalls and pyroclastic flows at Soufriere Hills Volcano, Montserrat. Memoirs 21(1):173-190; doi:10.1144/GSL.MEM.2002.021.01.08.

Cappello, A., G. Ganci, S. Calvari, N.M. Pérez, P.A. Hernández, S.V. Silva, J. Cabral, and C. Del Negro. 2016. Lava flow hazard modeling during the 2014-2015 Fogo eruption, Cape Verde. Journal of Geophysical Research: Solid Earth 121(4):2290-2303; doi:10.1002/2015JB012666.

Carazzo, G., E. Kaminski, and S. Tait. 2015. The timing and intensity of column collapse during explosive volcanic eruptions. Earth and Planetary Science Letters 411:208-217; doi:10.1016/j.epsl.2014.12.006.

Carey, R., V. Cayol, M. Poland, and D. Weis. 2015. Hawaiian Volcanoes: From Source to Surface. Hoboken, NJ: Wiley. doi:10.1002/9781118872079.

Carey, S., J. Gardner, and H. Sigurdsson. 1995. The intensity and magnitude of Holocene plinian eruptions from Mount St. Helens volcano. Journal of Volcanology and Geothermal Research 66(1):185-202; doi:10.1016/0377-0273(94)00059-P.

Caricchi, C., A. Vona, S. Corrado, G. Giordano, and C. Romano. 2014. 79AD Vesuvius PDC deposits' temperatures inferred from optical analysis on woods charred in-situ in the Villa dei Papiri at Herculaneum (Italy). Journal of Volcanology and Geothermal Research 289:14-25; doi:10.1016/j.jvolgeores.2014.10.016.

Carn, S.A., and F.J. Prata. 2010. Satellite-based constraints on explosive SO_2 release from Soufrière Hills Volcano, Montserrat. Geophysical Research Letters 37(19):L00E22; doi:10.1029/2010GL044971.

Carn, S.A., R.B. Watts, G. Thompson, and G.E. Norton. 2004. Anatomy of a lava dome collapse: The 20 March 2000 event at Soufrière Hills Volcano, Montserrat. Journal of Volcanology and Geothermal Research 131(3):241-264; doi:10.1016/S0377-0273(03)00364-0.

Carn, S.A., L. Clarisse, and A.J. Prata. 2016. Multi-decadal satellite measurements of global volcanic degassing. Journal of Volcanology and Geothermal Research 311:99-134; doi:10.1016/j.jvolgeores.2016.01.002.

Carr, B.B., A.B. Clarke, and L. Vanderkluysen. 2016. The 2006 lava dome eruption of Merapi Volcano (Indonesia): Detailed analysis using MODIS TIR. Journal of Volcanology and Geothermal Research 311:60-71; doi:10.1016/j.jvolgeores.2015.12.004.

Cas, R.A., and G. Giordano. 2014. Submarine volcanism: A review of the constraints, processes and products, and relevance to the Cabo de Gata volcanic succession. Italian Journal of Geosciences 133(3):362-377; doi:10.3301/IJG.2014.46.

Cashman, K.V., and G. Giordano. 2014. Calderas and magma reservoirs. Journal of Volcanology and Geothermal Research 288:28-45; doi:10.1016/j.jvolgeores.2014.09.007.

Cashman, K.V., and S.M. McConnell. 2005. Multiple levels of magma storage during the 1980 summer eruptions of Mount St. Helens, WA. Bulletin of Volcanology 68(1):57-75; doi:10.1007/s00445-005-0422-x.

Cashman, K.V., and A.C. Rust. 2016. Volcanic ash: Generation and spatial variations. Pp. 5-22 in Volcanic Ash: Hazard Observation, S. Mackie, K. Cashman, H. Ricketts, A. Rust, and M. Watson, eds. Amsterdam: Elsevier.

Cashman, K.V., and B. Scheu. 2015. Magmatic fragmentation. Pp. 459-471 in Encyclopedia of Volcanoes, 2nd Ed., H. Sigurdsson, B. Houghton, S.R. McNutt, H. Rymer, and J. Stix, eds. San Diego, CA: Academic Press. doi:10.1016/B978-0-12-385938-9.00025-0.

Cashman, K.V., S.A. Soule, B.H. Macket, N.I. Deligne, N.D. Deardorff, and H.R. Dietterich. 2013. How lava flows — New insights from applications of lidar technologies to lava flow studies. Geosphere 9(6):1664-1680; doi:10.1130/GES00706.

Cassidy, M., S. Watt, P. Talling, M. Palmer, M. Edmonds, M. Jutzeler, D. Wall-Palmer, M. Manga, M. Coussens, T. Gernon, R.N. Taylor, A. Michalik, E. Inglis, C. Breitkreuz, A. Le Friant, O. Ishizuka, G. Boudon, M.C. McCanta, T. Adachi, M.J. Hornbach, S.L. Colas, D. Endo, A. Fujinawa, K.S. Kataoka, F. Maeno, Y. Tamura, and F. Wang. 2015. Rapid onset of mafic magmatism facilitated by volcanic edifice collapse. Geophysical Research Letters 42(12):4778-4785; doi:10.1002/2015GL064519.

Castro, J.M., and D.B. Dingwell. 2009. Rapid ascent of rhyolitic magma at Chaitén volcano, Chile. Nature 461(7265):780-783; doi:10.1038/nature08458.

Castro, J.M., and J.E. Gardner. 2008. Did magma ascent rate control the explosive-effusive transition at the Inyo volcanic chain, California? Geology 36(4):279-282; doi:10.1130/G24453A.1.

Castro, J.M., B. Cordonnier, H. Tuffen, M.J. Tobin, L. Puskar, M.C. Martin, and H.A. Bechtel. 2012. The role of melt-fracture degassing in defusing explosive rhyolite eruptions at volcán Chaitén. Earth and Planetary Science Letters 333-334:63-69; doi:10.1016/j.epsl.2012.04.024.

Castruccio, A., A.C. Rust, and R.S.J. Sparks. 2013. Evolution of crust- and core-dominated lava flows using scaling analysis. Bulletin of Volcanology 75(1):681; doi:10.1007/s00445-012-0681-2.

Cembrano, J., and L. Lara. 2009. The link between volcanism and tectonics in the southern volcanic zone of the Chilean Andes: A review. Tectonophysics 471(1-2):96-113; doi:10.1016/j.tecto.2009.02.038.

Chadwick, W.W., D.J. Geist, S. Jonsson, M. Poland, D. Johnson, and C. Meertens. 2006. A volcano bursting at the seams: Inflation, faulting, and eruption at Sierra Negra volcano, Galapagos. Geology 34(12):1025-1028; doi:10.1130/G22826A.1.

Chadwick, W.W., K.V. Cashman, R.W. Embley, H. Matsumoto, R.P. Dziak, C.E.J. De Ronde, T.K. Lau, N.D. Deardorff, and S.G. Merle. 2008. Direct video and hydrophone observations of submarine explosive eruptions at NW Rota-1 volcano, Mariana arc. Journal of Geophysical Research: Solid Earth 113(B8):B08S10; doi:10.1029/2007JB005215.

Chadwick, W.W., J.B. Paduan, D.A. Clague, B.M. Dreyer, S.G. Merle, A.M. Bobbitt, D.W. Caress, B.T. Philip, D.S. Kelley, and S.L. Nooner. 2016. Voluminous eruption from a zoned magma body after an increase in supply rate at Axial Seamount. Geophysical Research Letters 43(23):12063-12070; doi:10.1002/2016GL071327.

Charbonnier, S.J., and R. Gertisser. 2008. Field observations and surface characteristics of pristine block-and-ash flow deposits from the 2006 eruption of Merapi Volcano, Java, Indonesia. Journal of Volcanology and Geothermal Research 177(4):971-982; doi:10.1016/j.jvolgeores.2008.07.008.

Chaussard, E., and F. Amelung. 2014. Regional controls on magma ascent and storage in volcanic arcs. Geochemistry, Geophysics, Geosystems 15(4):1407-1418; doi:10.1002/2013GC005216.

Chelle-Michou, C., B. Rottier, L. Caricchi, and G. Simpson. 2017. Tempo of magma degassing and the genesis of porphyry copper deposits. Scientific Reports 7:40,566; doi:10.1038/srep40566.

Chiodini, G., S. Caliro, P. De Martino, R. Avino, and F. Gherardi. 2012. Early signals of new volcanic unrest at Campi Flegrei caldera? Insights from geochemical data and physical simulations. Geology 40(10):943-946; doi:10.1130/G33251.1.

Chouet, B.A., and R.S. Matoza. 2013. A multi-decadal view of seismic methods for detecting precursors of magma movement and eruption. Journal of Volcanology and Geothermal Research 252:108-175; doi:10.1016/j.jvolgeores.2012.11.013.

Christiansen, R.L. 1984. Yellowstone magmatic evolution: Its bearing on understanding large-volume explosive volcanism. Pp. 84-95 in Explosive Volcanism: Inception, Evolution, and Hazards. Washington, DC: National Academy Press.

Christopher, T., M. Edmonds, M. Humphreys, and R.A. Herd. 2010. Volcanic gas emissions from Soufrière Hills Volcano, Montserrat 1995-2009, with implications for mafic magma supply and degassing. Geophysical Research Letters 37(19):L00E04; doi:10.1029/2009GL041325.

Christopher, T.E., J. Blundy, K. Cashman, P. Cole, M. Edmonds, P.J. Smith, R.S.J. Sparks, and A. Stinton. 2015. Crustal-scale degassing due to magma system destabilization and magma-gas decoupling at Soufrière Hills Volcano, Montserrat. Geochemistry, Geophysics, Geosystems 16(9):2797-2811; doi:10.1002/2015GC005791.

Cimarelli, C., A. Costa, S. Mueller, and H.M. Mader. 2011. Rheology of magmas with bimodal crystal size and shape distributions: Insights from analog experiments. Geochemistry, Geophysics, Geosystems 12(7); doi:10.1029/2011GC003606.

Clarke, A.B., J.C. Phillips, and K.N. Chojnicki. 2009. An investigation of Vulcanian eruption dynamics using laboratory analogue experiments and scaling analysis. Pp. 155-156 in Studies in Volcanology: The Legacy of George Walker, T. Thordarson, S. Self, G. Larsen, S.K. Rowland, and A. Hoskuldsson, eds. Special Publication of IAVCEI 2. London: Geological Society of London. doi:10.1144/IAVCEl002.

Clarke, D., F. Brenguier, J.L. Froger, N.M. Shapiro, A. Peltier, and T. Staudacher. 2013. Timing of a large volcanic flank movement at Piton de la Fournaise Volcano using noise-based seismic monitoring and ground deformation measurements. Geophysical Journal International 195(2):1132-1140; doi:10.1093/gji/ggt276.

Coffin, M.F., and O. Eldholm. 1994. Large igneous provinces: Crustal structure, dimensions, and external consequences. Reviews of Geophysics 32(1):1-36; doi:10.1029/93RG02508.

Coleman, D.S., J.M. Bartley, A.F. Glazner, and M.J. Pardue. 2012. Is chemical zonation in plutonic rocks driven by changes in source magma composition or shallow-crustal differentiation? Geosphere 8(6):1568-1587; doi:10.1130/Ges00798.1.

Comeau, M.J., M.J. Unsworth, and D. Cordell. 2016. New constraints on the magma distribution and composition beneath Volcán Uturuncu and the southern Bolivian Altiplano from magnetotelluric data. Geosphere 12(5):1391-1421; doi:10.1130/GES01277.1.

Cooper, K.M. 2015. Timescales of crustal magma reservoir processes: Insights from U-series crystal ages. Pp. 141-174 in Chemical, Physical and Temporal Evolution of Magmatic Systems, L. Caricchi and J.D. Blundy, eds. Special Publications 422. London: Geological Society of London. doi:10.1144/SP422.7.

Cooper, K.M., and A.J.R. Kent. 2014. Rapid remobilization of magmatic crystals kept in cold storage. Nature 506(7849):480-483; doi:10.1038/nature12991.

Costa, A., O. Melnik, and R.S.J. Sparks. 2007. Controls of conduit geometry and wallrock elasticity on lava dome eruptions. Earth and Planetary Science Letters 260(1-2):137-151; doi:10.1016/j.epsl.2007.05.024.

Costa, A., G. Wadge, and O. Melnik. 2012. Cyclic extrusion of a lava dome based on a stick-slip mechanism. Earth and Planetary Science Letters 337-338:39-46; doi:10.1016/j.epsl.2012.05.011.

Costa, A., Y.J. Suzuki, M. Cerminara, B.J. Devenish, T. Esposti Ongaro, M. Herzog, A.R. Van Eaton, L.C. Denby, M. Bursik, M. de'Michieli Vitturi, S. Engwell, A. Neri, S. Barsotti, A. Folch, G. Macedonio, F. Girault, G. Carazzo, S. Tait, E. Kaminski, L.G. Mastin, M.J. Woodhouse, J.C. Phillips, A.J. Hogg, W. Degruyter, and C. Bonadonna. 2016. Results of the eruptive column model inter-comparison study. Journal of Volcanology and Geothermal Research 326:2-25; doi:10.1016/j.jvolgeores.2016.01.017.

Costa, F., S. Chakraborty, and R. Dohmen. 2003. Diffusion coupling between trace and major elements and a model for calculation of magma residence times using plagioclase. Geochimica et Cosmochimica Acta 67(12):2189-2200; doi:10.1016/S0016-7037(02)01345-5.

Coussens, M., D. Wall-Palmer, P.J. Talling, S.F.L. Watt, M. Cassidy, M. Jutzeler, M.A. Clare, J.E. Hunt, M. Manga, T.M. Gernon, M.R. Palmer, S.J. Hatter, G. Boudon, D. Endo, A. Fujinawa, R. Hatfield, M.J. Hornbach, O. Ishizuka, K. Kataoka, A. Le Friant, F. Maeno, M. McCanta, and A.J. Stinton. 2016. The relationship between eruptive activity, flank collapse and sea-level at volcanic islands: A long-term (>1 Ma) record offshore, Lesser Antilles. Geochemistry, Geophysics, Geosystems 17(7):2591-2611; doi:10.1002/2015GC006053.

Crandell, D.R., and D.R. Mullineaux. 1978. Potential Hazards from Future Eruptions of Mount St. Helens Volcano, Washington. Geological Survey Bulletin No. 1383-C. Available at https://pubs.usgs.gov/bul/1383c/report.pdf. Accessed December 12, 2016.

Crews, J.B., and C.A. Cooper. 2014. Experimental evidence for seismically initiated gas bubble nucleation and growth in groundwater as a mechanism for coseismic borehole water level rise and remotely triggered seismicity. Journal of Geophysical Research: Solid Earth 119(9):7079-7091; doi:10.1002/2014JB011398.

Criswell, C.W. 1987. Chronology and pyroclastic stratigraphy of the May 18, 1980, eruption of Mount St. Helens, Washington. Journal of Geophysical Research: Solid Earth 92(B10):10237-10266; doi:10.1029/JB092iB10p10237.

Cronin, S.J., V.E. Neall, J.A. Lecointre, M.J. Hedley, and P. Loganathan. 2003. Environmental hazards of fluoride in volcanic ash: A case study from Ruapehu volcano, New Zealand. Journal of Volcanology and Geothermal Research 121(3-4):271-291; doi:10.1016/S0377-0273(02)00465-1.

Crowley, J.W., R.F. Katz, P. Huybers, C.H. Langmuir, and S.H. Park. 2015. Glacial cycles drive variations in the production of oceanic crust. Science 347(6227):1237-1240; doi:10.1126/science.1261508.

Cruz, F.G., and B.A. Chouet. 1997. Long-period events, the most characteristic seismicity accompanying the emplacement and extrusion of a lava dome in Galeras Volcano, Colombia, in 1991. Journal of Volcanology and Geothermal Research 77(1):121-158; doi:10.1016/S0377-0273(96)00091-1.

Dale, V.H., F.J. Swanson, and C.M. Crisafulli. 2005. Disturbance, survival, and succession: Understanding ecological responses to the 1980 eruption of Mount St. Helens. Pp. 3-11 in Ecological Responses to the 1980 Eruption of Mount St. Helens. New York: Springer.

Davis, J.L., L.H. Kellogg, J.R. Arrowsmith, B.A. Buffett, C.G. Constable, A. Donnellan, E.R. Ivins, G.S. Mattioli, S.E. Owen, M.E. Pritchard, M.E. Purucker, D.T. Sandwell, and J. Sauber. 2016. Challenges and Opportunities for Research in ESI (CORE): Report from the NASA Earth Surface and Interior (ESI) Focus Area Workshop, November 2-3, 2015, Arlington, VA. Available at https://smd-prod.s3.amazonaws.com/science-green/s3fs-public/atoms/files/CORE2016.pdf. Accessed March 1, 2017.

Day, S. 2015. Volcanic tsunamis. Pp. 993-1011 in Encyclopedia of Volcanoes, 2nd Ed., H. Sigurdsson, B. Houghton, S.R. McNutt, H. Rymer, and J. Stix, eds. San Diego, CA: Academic Press. doi:10.1016/B978-0-12-385938-9.00058-4.

De Siena, L., C. Thomas, G.P. Waite, S.C. Moran, and S. Klemme. 2014. Attenuation and scattering tomography of the deep plumbing system of Mount St. Helens. Journal of Geophysical Research: Solid Earth 119(11):8223-8238; doi:10.1002/2014JB011372.

Deeming, K.R., B. McGuire, and P. Harrop. 2010. Climate forcing of volcano lateral collapse: Evidence from Mount Etna, Sicily. Philosophical Transactions of the Royal Society A 368:2559-2577; doi:10.1098/rsta.2010.0054.

Degruyter, W., and C. Huber. 2014. A model for eruption frequency of upper crustal silicic magma chambers. Earth and Planetary Science Letters 403:117-130; doi:10.1016/j.epsl.2014.06.047.

Del Moral, R., and L.C. Bliss. 1993. Mechanisms of primary succession: Insights resulting from the eruption of Mount St Helens. Advances in Ecological Research 24:1-66; doi:10.1016/S0065-2504(08)60040-9.

Delle Donne, D.D., A.J.L. Harris, M. Ripepe, and R. Wright. 2010. Earthquake-induced thermal anomalies at active volcanoes. Geology 38(9):771-774; doi:10.1130/G30984.1.

Demouchy, S., S.D. Jacobsen, F. Gaillard, and C.R. Stern. 2006. Rapid magma ascent recorded by water diffusion profiles in mantle olivine. Geology 34(6):429-432; doi:10.1130/G22386.1.

Denlinger, R.P., and R.P. Hoblitt. 1999. Cyclic eruptive behavior of silicic volcanoes. Geology 27(5):459-462; doi:10.1130/0091-7613(1999)027<0459:CEBOSV>2.3.CO;2.

Desissa, M., N.E. Johnson, K.A. Whaler, S. Hautot, S. Fisseha, and G.J.K. Dawes. 2013. A mantle magma reservoir beneath an incipient mid-ocean ridge in Afar, Ethiopia. Nature Geoscience 6(10):861-865; doi:10.1038/ngeo1925.

Di Muro, A., A. Neri, and M. Rosi. 2004. Contemporaneous convective and collapsing eruptive dynamics: The transitional regime of explosive eruptions. Geophysical Research Letters 31(10):L10607; doi:10.1029/2004GL019709.

Dietterich, H.R., and K.V. Cashman. 2014. Channel networks within lava flows: Formation, evolution, and implications for flow behavior. Journal of Geophysical Research: Earth Surface 119(8):1704-1724; doi:10.1002/2014JF003103.

Dingwell, D.B. 1996. Volcanic dilemma: Flow or blow? Science 273(5278):1054-1055; doi:10.1126/science.273.5278.1054.

Dmitrieva, K., A.J. Hotovec-Ellis, S. Prejean, and E.M. Dunham. 2013. Frictional-faulting model for harmonic tremor before Redoubt Volcano eruptions. Nature Geoscience 6(8):652-656; doi:10.1038/ngeo1879.

Druitt, T.H., S.R. Young, B. Baptie, C. Bonadonna, E.S. Calder, A.B. Clarke, P.B. Cole, C.L. Harford, R.A. Herd, R. Luckett, G. Ryan, and B. Voight. 2002. Episodes of cyclic Vulcanian explosive activity with fountain collapse at Soufrière Hills Volcano, Montserrat. Memoirs 21:281-306; doi:10.1144/GSL.MEM.2002.021.01.13.

Druitt, T.H., F. Costa, E. Deloule, M. Dungan, and B. Scaillet. 2012. Decadal to monthly timescales of magma transfer and reservoir growth at a caldera volcano. Nature 482(7383):77-80; doi:10.1038/nature10706.

Dufek, J. 2016. The fluid mechanics of pyroclastic density currents. Annual Review of Fluid Mechanics 48:459-485; doi:10.1146/annurev-fluid-122414-034252.

Dufek, J., and M. Manga. 2008. The in-situ production of ash in pyroclastic flows. Journal of Geophysical Research: Solid Earth 113:B09207; doi:10.1029/2007JB005555.

Dufek, J., M. Manga, and M. Staedter. 2007. Littoral blasts: Pumice-water heat transfer and the conditions for steam explosions when pyroclastic flows enter the ocean. Journal of Geophysical Research: Solid Earth 112:B11201; doi:10.1029/2006JB004910.

Dufek, J., M. Manga, and A. Patel. 2012. Granular disruption during explosive volcanic eruptions. Nature Geoscience 5(8):561-564; doi:10.1038/ngeo1524.

Duggen, S., N. Olgun, P. Croot, L. Hoffmann, H. Dietze, P. Delmelle, and C. Teschner. 2010. The role of airborne volcanic ash for the surface ocean biogeochemical iron-cycle: A review. Biogeosciences 7(3):827-844; doi:10.5194/bg-7-827-2010.

Dull, R.A., J.R. Southon, and P. Sheets. 2001. Volcanism, ecology and culture: A reassessment of the Volcán Ilopango TBJ eruption in the southern Maya realm. Latin American Antiquity 12(1):25-44; doi:10.2307/971755.

Dunbar, N. 1999. Cosmogenic ^{36}Cl-deteremined age of the Carrizozo lava flows, south-central New Mexico. New Mexico Geology 21(2):25-29.

Durant, A.J., W.I. Rose, A.M. Sarna-Wojcicki, S. Carey, and A.C.M. Volentik. 2009. Hydrometeor-enhanced tephra sedimentation: Constraints from the 18 May 1980 eruption of Mount St. Helens. Journal of Geophysical Research: Solid Earth 114(B3):B03204; doi:10.1029/2008JB005756.

Ebinger, C., A. Ayele, D. Keir, J. Rowland, G. Yirgu, T. Wright, M. Belachew, and I. Hamling. 2010. Length and timescales of rift faulting and magma intrusion: The Afar rifting cycle from 2005 to present. Annual Review of Earth and Planetary Sciences 38:439-466; doi:10.1146/annurev-earth-040809-152333.

Ebmeier, S.K., J.R. Elliott, J.M. Nocquet, J. Biggs, P. Mothes, P. Jarrín, P. Yépez, S. Aguaiza, P. Lundgren, and S.V. Samsonov. 2016. Shallow earthquake inhibits unrest near Chiles–Cerro Negro volcanoes, Ecuador–Colombian border. Earth and Planetary Science Letters 450:283-291; doi:10.1016/j.epsl.2016.06.046.

Eckhardt, S., A.J. Prata, P. Seibert, K. Stebel, and A. Stohl. 2008. Estimation of the vertical profile of sulfur dioxide injection into the atmosphere by a volcanic eruption using satellite column measurements and inverse transport modeling. Atmospheric Chemistry and Physics 8(14):3881-3897; doi:10.5194/acp-8-3881-2008.

Edmonds, M., D. Pyle, and C. Oppenheimer. 2001. A model for degassing at the Soufrière Hills Volcano, Montserrat, West Indies, based on geochemical data. Earth and Planetary Science Letters 186(2):159-173; doi:10.1016/S0012-821X(01)00242-4.

Egan, J., W.J. Fletcher, T.E. Allott, C.S. Lane, J.J. Blackford, and D.H. Clark. 2016. The impact and significance of tephra deposition on a Holocene forest environment in the North Cascades, Washington, USA. Quaternary Science Reviews 137:135-155; doi:10.1016/j.quascirev.2016.02.013.

Eichelberger, J.C. 1997. Drilling volcanoes. Science 278(5340): 1084-1085; doi:10.1126/science.278.5340.1084.

Eichelberger, J.C., P.C. Lysne, and L.W. Younker. 1984. Research drilling at Inyo Domes, Long Valley Caldera, California. EOS 65(39):721-725; doi:10.1029/EO065i039p00721.

Elders, W.A., G.O. Friðleifsson, R.A. Zierenberg, E.C. Pope, A.K. Mortensen, A. Guðmundsson, J.B. Lowenstern, N.E. Marks, L. Owens, D.K. Bird, M. Reed, N.J. Olsen, and P.A. Schiffman. 2011. Origin of a rhyolite that intruded a geothermal well while drilling in a basaltic volcano, at Krafla, Iceland. Geology 39(3):231-234; doi:10.1130/G31393.1.

Elders, W.A., G.O. Friðleifsson, and A. Albertsson. 2014. Drilling into magma and the implications of the Iceland Deep Drilling Project (IDDP) for high-temperature geothermal systems worldwide. Geothermics 49:111-118; doi:10.1016/j.geothermics.2013.05.001.

Elsworth, D., B. Voight, G. Thompson, and S.R. Young. 2004. Thermal-hydrologic mechanism for rainfall-triggered collapse of lava domes. Geology 32(11):969-972; doi:10.1130/G20730.1.

Endo, E.T., and T. Murray. 1991. Real-time seismic amplitude measurement (RSAM): A volcano monitoring and prediction tool. Bulletin of Volcanology 53(7):533-545; doi:10.1007/BF00298154.

Esposti-Ongaro, T., A.B. Clarke, B. Voight, A. Neri, and C. Widiwijayanti. 2012. Multiphase flow dynamics of pyroclastic density currents during the May 18, 1980 lateral blast of Mount St. Helens (USA). Journal of Geophysical Research: Solid Earth 117(B6):B06208; doi:10.1029/2011JB009081.

Ewert, J.W., M. Guffanti, and T.L. Murray. 2005. An Assessment of Volcanic Threat and Monitoring Capabilities in the United States: Framework for a National Volcano Early Warning System. Open-File Report No. 2005-1164. U.S. Geological Survey. Available at https://pubs.usgs.gov/of/2005/1164/2005-1164.pdf. Accessed December 12, 2016.

Eychenne, J., J.L. Le Pennec, L. Troncoso, M. Gouhier, and J.M. Nedelec. 2012. Causes and consequences of bimodal grain-size distribution of tephra fall deposited during the August 2006 Tungurahua eruption (Ecuador). Bulletin of Volcanology 74(1):187-205; doi:10.1007/s00445-011-0517-5.

Eychenne, J., J.L. Le Pennec, P. Ramón, and H. Yepes. 2013. Dynamics of explosive paroxysms at open-vent andesitic systems: High-resolution mass distribution analyses of the 2006 Tungurahua fall deposit (Ecuador). Earth and Planetary Science Letters 361:343-355; doi:10.1016/j.epsl.2012.11.002.

Fauria, K.E., M. Manga, and M. Chamberlain. 2016. Effect of particle entrainment on the runout of pyroclastic density currents. Journal of Geophysical Research: Solid Earth 121(9):6445-6461; doi:10.1002/2016JB013263.

Favalli, M., S. Tarquini, A. Fornaciai, and E. Boschi. 2009. A new approach to risk assessment of lava flow at Mount Etna. Geology 37(12):1111-1114; doi:10.1130/G30187A.1.

Fee, D., M. Garces, and A. Steffke. 2010. Infrasound from Tungurahua Volcano 2006-2008: Strombolian to Plinian eruptive activity. Journal of Volcanology and Geothermal Research 193(1-2):67-81; doi:10.1016/j.jvolgeores.2010.03.006.

Fenton, C. 2012. Age-dating of young basalts: A cosmogenic ^{3}He and ^{21}Ne study at Sunset Crater and the SP Flow, AZ, USA. Quaternary International 279-280:139; doi:10.1016/j.quaint.2012.08.091.

Feuillet, N., F. Beauducel, and P. Tapponnier. 2011. Tectonic context of moderate to large historical earthquakes in the Lesser Antilles and mechanical coupling with volcanoes. Journal of Geophysical Research: Solid Earth 116(B10):B10308; doi:10.1029/2011JB008443.

Fink, J.H., and S.W. Kieffer. 1993. Estimate of pyroclastic flow velocities resulting from explosive decompression of lava domes. Nature 363(6430):612-615; doi:10.1038/363612a0.

Fioletov, V.E., C.A. McLinden, N. Krotkov, C. Li, J. Joiner, N. Theys, S. Carn, and M.S. Moran. 2016. A global catalogue of large SO_2 sources and emissions derived from the Ozone Monitoring Instrument. Atmospheric Chemistry and Physics 16(18):11497-11519; doi:10.5194/acp-16-11497-2016.

Fischer, T.P., M.M. Morrissey, V.M.L. Calvache, M.D. Gomez, C.R. Torres, J. Stix, and S.N. Williams. 1994. Correlations between SO_2 flux and long-period seismicity at Galeras volcano. Nature 368(6467):135-137; doi:10.1038/368135a0.

Fisher, R.V., G. Orsi, M. Ort, and G. Heiken. 1993. Mobility of a large-volume pyroclastic flow-emplacement of the Campanian ignimbrite, Italy. Journal of Volcanology and Geothermal Research 56(3):205-220; doi:10.1016/0377-0273(93)90017-L.

Fiske, R.S. 1963. Subaqueous pyroclastic flows in the Ohanapecosh Formation, Washington. Geological Society of America Bulletin 74(4):391-406; doi:10.1130/0016-7606(1963)74[391:SPFITO]2.0.CO;2.

Fiske, R.S., K.V. Cashman, A. Shibata, and K. Watanabe. 1998. Tephra dispersal from Myojinsho, Japan, during its shallow submarine eruption of 1952–1953. Bulletin of Volcanology 59(4):262-275; doi:10.1007/s004450050190.

Fitton, J.G., D. James, and W.P. Leeman. 1991. Basic magmatism associated with late Cenozoic extension in the western United States: Compositional variations in space and time. Journal of Geophysical Research: Solid Earth 96(B8):13693-13771; doi:10.1029/91JB00372.

Formenti, Y., T.H. Druitt, and K. Kelfoun. 2003. Characterisation of the 1997 Vulcanian explosions of Soufrière Hills Volcano, Montserrat, by video analysis. Bulletin of Volcanology 65(8):587-605; doi:10.1007/s00445-003-0288-8.

Fournier, T.J., M.E. Pritchard, and S.N. Riddick. 2010. Duration, magnitude, and frequency of subaerial volcano deformation events: New results from Latin America using InSAR and a global synthesis. Geochemistry, Geophysics, Geosystems 11(1):Q01003; doi:10.1029/2009GC002558.

Friðleifsson, G.Ó., W.A. Elders, G. Bignall, and D. Nielson. 2013. A plan for a 5 km-deep borehole at Reykjanes, Iceland, into the root zone of a black smoker on land. Scientific Drilling 16:73-79; doi:10.5194/sd-16-73-2013.

Fujita, E., T. Kozono, H. Ueda, Y. Kohno, S. Yoshioka, N. Toda, A. Kikuchi, and Y. Ida. 2013. Stress field change around the Mount Fuji volcano magma system caused by the Tohoku megathrust earthquake, Japan. Bulletin of Volcanology 75(1):679; doi:10.1007/s00445-012-0679-9.

Funiciello, R., G. Giordano, and D. De Rita. 2003. The Albano maar lake (Colli Albani Volcano, Italy): Recent volcanic activity and evidence of pre-Roman Age catastrophic lahar events. Journal of Volcanology and Geothermal Research 123(1-2):43-61; doi:10.1016/S0377-0273(03)00027-1.

Gaunt, H.E., P.R. Sammonds, P.G. Meredith, R. Smith, and J.S. Pallister. 2014. Pathways for degassing during the lava dome eruption of Mount St. Helens 2004-2008. Geology 42(11):947-950; doi:10.1130/G35940.1.

Girona, T., F. Costa, and G. Schubert. 2015. Degassing during quiescence as a trigger of magma ascent and volcanic eruptions. Scientific Reports 5:18212; doi:10.1038/srep18212.

Gíslason, S.R., G. Stefánsdóttir, M.A. Pfeffer, S. Barsotti, T. Jóhannsson, I. Galeczka, E. Bali, O. Sigmarsson, A. Stefánsson, N.S. Keller, A. Sigurdsson, B. Bergsson, B. Galle, V.C. Jacobo, S. Arellano, A. Aiuppa, E.B. Jónasdóttir, E.S. Eiríksdóttir, S. Jakobsson, G.H. Guðfinnsson, S.A. Halldórsson, H. Gunnarsson, B. Haddadi, I. Jónsdóttir, T. Thordarson, M. Riishuus, T. Högnadóttir, T. Dürig, G.B.M. Pedersen, Á. Höskuldsson, and M.T. Gudmundsson. 2015. Environmental pressure from the 2014-15 eruption of Bárðarbunga volcano, Iceland. Geochemical Perspectives Letters 1(1):84-93; doi:10.7185/geochemlet.1509.

Glaze, L.S., S.M. Baloga, and L. Wilson. 1997. Transport of atmospheric water vapor by volcanic eruption columns. Journal of Geophysical Research: Atmospheres 102(D5):6099-6108; doi:10.1029/96JD03125.

Gonnermann, H.M. 2015. Magma fragmentation. Annual Review of Earth and Planetary Sciences 43:431-458; doi:10.1146/annurev-earth-060614-105206.

Gonnermann, H.M., and M. Manga. 2003. Explosive volcanism may not be an inevitable consequence of magma fragmentation. Nature 426(6965):432-435; doi:10.1038/nature02138.

Gonnermann, H.M., and M. Manga. 2012. Dynamics of magma ascent in the volcanic conduit. Pp. 55-84 in Modeling Volcanic Processes: The Physics and Mathematics of Volcanism, S.A. Fagents, T.K.P. Gregg, and R.M.C. Lopes, eds. New York: Cambridge University Press. Available at http://seismo.berkeley.edu/~manga/gonnermannmanga2012.pdf. Accessed December 12, 2016.

Gouhier, M., and F. Donnadieu. 2010. The geometry of Strombolian explosions: Insights from Doppler radar measurements. Geophysical Journal International 183(3):1376-1391; doi:10.1111/j.1365-246X.2010.04829.x.

Greenfield, T., R.S. White, and S. Roecker. 2016. The magmatic plumbing system of the Askja central volcano, Iceland, as imaged by seismic tomography. Journal of Geophysical Research: Solid Earth 121(10):7211-7229; doi:10.1002/2016JB013163.

Griffiths, R.W., and J.H. Fink. 1997. Solidifying Bingham extrusions: A model for the growth of silicic lava domes. Journal of Fluid Mechanics 347:13-36; doi:10.1017/S0022112097006344.

Gudmundsson, A. 2012. Magma chambers: Formation, local stresses, excess pressures, and compartments. Journal of Volcanology and Geothermal Research 237:19-41; doi:10.1016/j.jvolgeores.2012.05.015.

Guevara-Murua, A., E.J. Hendy, A.C. Rust, and K.V. Cashman. 2015. Consistent decrease in North Atlantic Tropical Cyclone frequency following major volcanic eruptions in the last three centuries. Geophysical Research Letters 42(21):9425-9432; doi:10.1002/2015GL066154.

Guffanti, M., T.J. Casadevall, and K. Budding. 2010. Encounters of Aircraft with Volcanic Ash Clouds; A Compilation of Known Incidents, 1953–2009. Data Series 545, Version 1.0. Reston, VA: U.S. Geological Survey. Available at https://pubs.usgs.gov/ds/545/DS545.pdf. Accessed March 1, 2017.

Gunnarsson, T.G., L. Jóhannesdóttir, J.A. Alves, B. Þórisson, and J.A. Gill. 2017. Effects of spring temperature and volcanic eruptions on wader productivity. Ibis 159(2):467-471; doi:10.1111/ibi.12449.

Hammer, J.E. 2008. Experimental studies of the kinetics and energetics of magma crystallization. Pp. 9-60 in Minerals, Inclusions and Volcanic Processes, K. Putirka and F.J. Tepley III, eds. Reviews in Mineralogy and Geochemistry Vol. 69. Chantilly, VA: Mineralogical Society of America.

Harris, A., T. De Groeve, S. Carn, and F. Garel. 2016. Risk evaluation, detection and simulation during effusive eruption disasters. Pp. 1-22 in Detecting, Modelling and Responding to Effusive Eruptions. Special Publications 426. Geological Society of London. doi:10.1144/SP426.29.

Harris, A.J.L., and M. Ripepe. 2007. Regional earthquake as a trigger for enhanced volcanic activity: Evidence from MODIS thermal data. Geophysical Research Letters 34(2):L02304; doi:10.1029/2006GL028251.

Harrison, M., S. Baldwin, M. Caffee, G. Gehrels, B. Schoene, D. Shuster, and B. Singer. 2015. Geochronology: It's about time. EOS Opinions, December 28; doi:10.1029/2015EO041901.

Heliker, C., and T.N. Mattox. 2003. The first two decades of the Pu'u 'Ō'ō-Kūpaianaha eruption: Chronology and selected bibliography. Pp. 1-28 in The Pu'u 'Ō'ō-Kūpaianaha Eruption of Kīlauea Volcano, Hawai'i: The First 20 Years, C. Heliker, D.A. Swanson, and T.J. Takahashi, eds. U.S. Geological Survey Professional Paper 1676. Available at https://pubs.usgs.gov/pp/pp1676. Accessed December 13, 2016.

Hildreth, W. 2007. Quaternary Magmatism in the Cascades-Geologic Perspectives. U.S. Geological Survey Professional Paper 1744. Available at https://pubs.usgs.gov/pp/pp1744/pp1744.pdf. Accessed December 13, 2016.

Hildreth, W., and J. Fierstein. 2012. The Novarupta-Katmai Eruption of 1912–Largest Eruption of the Twentieth Century; Centennial Perspectives. U.S. Geological Survey Professional Paper 1791. Reston, VA: U.S. Geological Survey. Available at https://pubs.usgs.gov/pp/1791/pp1791.pdf. Accessed March 1, 2017.

Hildreth, W., J. Fierstein, and A.T. Calvert. 2012. Geologic Map of Three Sisters Volcanic Cluster, Cascade Range, Oregon. U.S. Geological Survey Scientific Investigations Map 3186. Available at https://pubs.usgs.gov/sim/3186. Accessed December 13, 2016.

Hill, D.P., F. Pollitz, and C. Newhall. 2002. Earthquake-volcano interactions. Physics Today 55(11):41-47; doi: 10.1063/1.1535006.

Hoblitt, R.P., C.D. Miller, and J.W. Vallance. 1981. Origin and stratigraphy of the deposit produced by the May 18 directed blast. Pp. 401-420 in The 1980 Eruptions of Mount St. Helens, Washington, P.W. Lipman and D.R. Mullineux, eds. U.S. Geological Survey Professional Paper 1250. Available at https://pubs.usgs.gov/pp/1250/report.pdf. Accessed December 13, 2016.

Horwell, C.J., P.J. Baxter, and R. Kamanyire. 2015. Health impacts of volcanic eruptions. Pp. 289-294 in Global Volcanic Hazards and Risk, S.C. Loughlin, S. Sparks, S.K. Brown, S.F. Jenkins, and C. Vye-Brown, eds. Cambridge, UK: Cambridge University Press. doi:https://doi.org/10.1017/CBO9781316276273.015.

Höskuldsson, Á., N. Óskarsson, R. Pedersen, K. Grönvold, K. Vogfjörð, and R. Ólafsdóttir. 2007. The millennium eruption of Hekla in February 2000. Bulletin of Volcanology 70(2):169-182; doi:10.1007/s00445-007-0128-3.

Hotovec, A.J., S.G. Prejean, J.E. Vidale, and J. Gomberg. 2013. Strongly gliding harmonic tremor during the 2009 eruption of Redoubt Volcano. Journal of Volcanology and Geothermal Research 259:89-99; doi:10.1016/j.jvolgeores.2012.01.001.

Houghton, B.F., and H.M. Gonnermann. 2008. Basaltic explosive volcanism: Constraints from deposits and models. Chemie der Erde 68(2):117-140; doi:10.1016/j.chemer.2008.04.002.

Houghton, B.F., J.D. White, and A.M. Van Eaton. 2015. Phreatomagmatic and related eruption styles. Pp. 537-552 in Encyclopedia of Volcanoes, 2nd Ed., H. Sigurdsson, B. Houghton, S.R. McNutt, H. Rymer, and J. Stix, eds. San Diego, CA: Academic Press. doi:10.1016/B978-0-12-385938-9.00030-4.

Huang, H.H., F.C. Lin, B. Schmandt, J. Farrell, R.B. Smith, and V.C. Tsai. 2015. The Yellowstone magmatic system from the mantle plume to the upper crust. Science 348(6236):773-776; doi:10.1126/science.aaa5648.

Huber, C., O. Bachmann, and M. Manga. 2010. Two competing effects of volatiles on heat transfer in crystal-rich magmas: Thermal insulation vs defrosting. Journal of Petrology 51(4):847-867; doi:10.1093/petrology/egq003.

Hughes, E.J., J. Yorks, N.A. Krotkov, A.M. da Silva, and M. McGill. 2016. Using CATS near-real-time lidar observations to monitor and constrain volcanic sulfur dioxide (SO_2) forecasts. Geophysical Research Letters 43(20):11089-11097; doi:10.1002/2016GL070119.

Humphreys, M.C., T. Menand, J.D. Blundy, and K. Klimm. 2008. Magma ascent rates in explosive eruptions: Constraints from H_2O diffusion in melt inclusions. Earth and Planetary Science Letters 270(1-2):25-40; doi:10.1016/j.epsl.2008.02.041.

Hurwitz, S., and M. Manga. 2017. The fascinating and complex dynamics of geyser eruptions. Annual Reviews of Earth and Planetary Science 45; doi:10.1146/annurev-earth-063016-015605.

Huybers, P., and C. Langmuir. 2017. Delayed CO_2 emissions from mid-ocean ridge volcanism as a possible cause of late-Pleistocene glacial cycles. Earth and Planetary Science Letters 457:238-249; doi:10.1016/j.epsl.2016.09.021.

Iacovino, K. 2015. Linking subsurface to surface degassing at active volcanoes: A thermodynamic model with applications to Erebus volcano. Earth and Planetary Science Letters 431:59-74; doi:10.1016/j.epsl.2015.09.016.

IAVCEI (International Association of Volcanology and Chemistry of the Earth's Interior) Task Group on Crisis Protocols. 2016. Toward IAVCEI guidelines on the roles and responsibilities of scientists involved in volcanic hazard evaluation, risk mitigation, and crisis response. Bulletin of Volcanology 78(4):31; doi:10.1007/s00445-016-1021-8.

Ingebritsen, S.E., D.R. Shelly, P.A. Hsieh, L.E. Clor, P.H. Seward, and W.C. Evans. 2015. Hydrothermal response to a volcano-tectonic earthquake swarm, Lassen, California. Geophysical Research Letters 42(21):9223-9230; doi:10.1002/2015GL065826.

Ishihara, K. 1985. Dynamical analysis of volcanic explosion. Journal of Geodynamics 3(3-4):327-349; doi:10.1016/0264-3707(85)90041-9.

Iverson, R.M., S.P. Schilling, and J.W. Vallance. 1998. Objective delineation of lahar-inundation hazard zones. Geological Society of America Bulletin 110(8):972-984; doi:10.1130/0016-7606(1998)110<0972:ODOLIH>2.3.CO;2.

Iverson, R.M., D. Dzurisin, C.A. Gardner, T.M. Gerlach, R.G. LaHusen, M. Lisowski, J.J. Major, S.D. Malone, J.A. Messerich, S.C. Moran, J.S. Pallister, A.I. Qamar, S.P. Schilling, and J.W. Vallance. 2006. Dynamics of seismogenic volcanic extrusion at Mount St Helens in 2004-05. Nature 444(7118):439-443; doi:10.1038/nature05322.

James, M.R., H. Pinkerton, and S. Robson. 2007. Image-based measurement of flux variation in distal regions of active lava flows. Geochemistry, Geophysics, Geosystems 8(3):Q03006; doi:10.1029/2006GC001448.

James, M.R., H. Pinkerton, and M. Ripepe. 2010. Imaging short period variations in lava flux. Bulletin of Volcanology 72(6):671-676; doi:10.1007/s00445-010-0354-y.

Jaupart, C., and C.J. Allègre. 1991. Gas content, eruption rate and instabilities of eruption regime in silicic volcanoes. Earth and Planetary Science Letters 102(3-4):413-429; doi:10.1016/0012-821X(91)90032-D.

Jaxybulatov, K., N.M. Shapiro, I. Koulakov, A. Mordret, M. Landès, and C. Sens-Schönfelder. 2014. A large magmatic sill complex beneath the Toba caldera. Science 346(6209):617-619; doi:10.1126/science.1258582.

Jellinek, A.M., M. Manga, and M.O. Saar. 2004. Did melting glaciers cause volcanic eruptions in eastern California? Probing the mechanics of dike formation. Journal of Geophysical Research: Solid Earth 109(B9):B09206; doi:10.1029/2004JB002978.

Jenkins, S., C. Magill, J. McAneney, and R. Blong. 2012. Regional ash fall hazard I: A probabilistic assessment methodology. Bulletin of Volcanology 74(7):1699-1712; doi:10.1007/s00445-012-0627-8.

Johnson, J.B., and J.M. Lees. 2000. Plugs and chugs—seismic and acoustic observations of degassing explosions at Karymsky, Russia and Sangay, Ecuador. Journal of Volcanology and Geothermal Research 101(1):67-82; doi:10.1016/S0377-0273(00)00164-5.

Johnson, J.B., J.J. Lyons, B.J. Andrews, and J.M. Lees. 2014. Explosive dome eruptions modulated by periodic gas-driven inflation. Geophysical Research Letters 41(19):6689-6697; doi:10.1002/2014GL061310.

Jutzeler, M., R. Marsh, R.J. Carey, J.D. White, P.J. Talling, and L. Karlstrom. 2014. On the fate of pumice rafts formed during the 2012 Havre submarine eruption. Nature Communications 5:3660; doi:10.1038/ncomms4660.

Kahl, M., S. Chakraborty, F. Costa, M. Pompilio, M. Liuzzo, and M. Viccaro. 2013. Compositionally zoned crystals and real-time degassing data reveal changes in magma transfer dynamics during the 2006 summit eruptive episodes of Mt. Etna. Bulletin of Volcanology 75(2):692; doi:10.1007/s00445-013-0692-7.

Kaiser, J.F., S. de Silva, A.K. Schmitt, R. Economos, and M. Sunagua. 2017. Million-year melt—presence in monotonous intermediate magma for a volcanic–plutonic assemblage in the Central Andes: Contrasting histories of crystal-rich and crystal-poor super-sized silicic magmas. Earth and Planetary Science Letters 457:73-86; doi:10.1016/j.epsl.2016.09.048.

Kaneko, T., F. Maeno, and S. Nakada. 2016. 2014 Mount Ontake eruption: Characteristics of the phreatic eruption as inferred from aerial observations. Earth, Planets and Space 68:72; doi:10.1186/s40623-016-0452-y.

Karlstrom, L., and M. Richards. 2011. On the evolution of large ultramafic magma chambers and timescales for flood basalt eruptions. Journal of Geophysical Research: Solid Earth 116(B8); doi:10.1029/2010JB008159.

Karlstrom, L., J. Dufek, and M. Manga. 2009. Organization of volcanic plumbing through magmatic lensing by magma chambers and volcanic loads. Journal of Geophysical Research: Solid Earth 114(B10):B10204; doi:10.1029/2009JB006339.

Kato, A., T. Terakawa, Y. Yamanaka, Y. Maeda, S. Horikawa, K. Matsuhiro, and T. Okuda. 2015. Preparatory and precursory processes leading up to the 2014 phreatic eruption of Mount Ontake, Japan. Earth, Planets and Space 67:111; doi:10.1186/s40623-015-0288-x.

Kauahikaua, J., M. Mangan, C. Heliker, and T. Mattox. 1996. A quantitative look at the demise of a basaltic vent: The death of Kupaianaha, Kilauea Volcano, Hawai'i. Bulletin of Volcanology 57(8):641-648; doi:10.1007/s004450050117.

Kelemen, P.B., and C.E. Manning. 2015. Reevaluating carbon fluxes in subduction zones, what goes down, mostly comes up. Proceedings of the National Academy of Sciences of the United States of America 112(30):E3997-E4006; doi:10.1073/pnas.1507889112.

Keller, G.V., L.T. Grose, J.C. Murray, and C.K. Skokan. 1979. Results of an experimental drill hole at the summit of Kilauea volcano, Hawaii. Journal of Volcanology and Geothermal Research 5(3-4):345-385; doi:10.1016/0377-0273(79)90024-6.

Kennedy, B., O. Spieler, B. Scheu, U. Kueppers, J. Taddeucci, and D.B. Dingwell. 2005. Conduit implosion during Vulcanian eruptions. Geology 33(7):581-584; doi:10.1130/G21488.1.

Kent, A.J.R. 2008. Melt inclusions in basaltic and related volcanic rocks. Pp. 273-331 in Minerals, Inclusions and Volcanic Processes, K. Putirka and F.J. Tepley III, eds. Reviews in Mineralogy and Geochemistry Vol. 69. Chantilly, VA: Mineralogical Society of America.

Kieffer, S.W. 1981. Fluid dynamics of the May 18 blast at Mount St. Helens. Pp. 379-400 in The 1980 Eruptions of Mount St. Helens, Washington, P.W. Lipman and D.R. Mullineux, eds. U.S. Geological Survey Professional Paper 1250. Available at https://pubs.usgs.gov/pp/1250/report.pdf. Accessed December 13, 2016.

Kiser, E., A. Levander, S. Harder, G. Abers, K. Creager, J. Vidale, S. Moran, and S. Malone. 2013. iMUSH: The design of the Mount St. Helens high-resolution active source seismic experiment. Abstract EGU2013-12325. European Geosciences Union General Assembly 2013, April 7-12, 2013, Vienna, Austria. Available at http://meetingorganizer.copernicus.org/EGU2013/EGU2013-12325.pdf. Accessed December 13, 2016.

Kiyosugi, K., C. Connor, R.S.J. Sparks, H.S. Crosweller, S.K. Brown, L. Siebert, T. Wang, and S. Takarada. 2015. How many explosive eruptions are missing from the geologic record? Analysis of the Quaternary record of large magnitude explosive eruptions in Japan. Journal of Applied Volcanology 4:17; doi:10.1186/s13617-015-0035-9.

Kokelaar, B.P., and G.P. Durant. 1983. The submarine eruption and erosion of Surtla (Surtsey), Iceland. Journal of Volcanology and Geothermal Research 19(3-4):239-246; doi:10.1016/0377-0273(83)90112-9.

Komorowski, J.C., S. Jenkins, P.J. Baxter, A. Picquout, F. Lavigne, S. Charbonnier, R. Gertisser, K. Preece, N. Cholik, A. Budi-Santoso, and S. Surono. 2013. Paroxysmal dome explosion during the Merapi 2010 eruption: Processes and facies relationships of associated high-energy pyroclastic density currents. Journal of Volcanology and Geothermal Research 261:260-294; doi:10.1016/j.jvolgeores.2013.01.007.

Kravitz, B., and A. Robock. 2011. Climate effects of high-latitude volcanic eruptions: Role of the time of year. Journal of Geophysical Research: Atmospheres 116(D1):D01105; doi:10.1029/2010JD014448.

Kristiansen, N.I., A.J. Prata, A. Stohl, and S.A. Carn. 2015. Stratospheric volcanic ash emissions from the 13 February 2014 Kelut eruption. Geophysical Research Letters 42(2):588-596; doi:10.1002/2014GL062307.

Krotkov, N.A., M.R. Schoeberl, G.A. Morris, S. Carn, and K. Yang. 2010. Dispersion and lifetime of the SO_2 cloud from the August 2008 Kasatochi eruption. Journal of Geophysical Research: Atmospheres 115(D2):D00L20; doi:10.1029/2010JD013984.

Kubanek, J., J.A. Richardson, S.J. Charbonnier, and L.J. Connor. 2015. Lava flow mapping and volume calculations for the 2012–2013 Tolbachik, Kamchatka, fissure eruption using bistatic TanDEM-X InSAR. Bulletin of Volcanology 77:106; doi:10.1007/s00445-015-0989-9.

Kueppers, U., B. Scheu, O. Spieler, and D.B. Dingwell. 2006. Fragmentation efficiency of explosive volcanic eruptions: A study of experimentally generated pyroclasts. Journal of Volcanology and Geothermal Research 153(1-2):125-135; doi:10.1016/j.jvolgeores.2005.08.006.

Kutterolf, S., T.H. Hansteen, K. Appel, A. Freundt, K. Krüger, W. Pérez, and H. Wehrmann. 2013. Combined bromine and chlorine release from large explosive volcanic eruptions: A threat to stratospheric ozone? Geology 41(6):707-710; doi:10.1130/G34044.1.

La Femina, P.C., C.B. Connor, B.E. Hill, W. Strauch, and J.A. Saballos. 2004. Magma–tectonic interactions in Nicaragua: The 1999 seismic swarm and eruption of Cerro Negro volcano. Journal of Volcanology and Geothermal Research 137(1-3):187-199; doi:10.1016/j.jvolgeores.2004.05.006.

Langmann, B., K. Zakšek, M. Hort, and S. Duggen. 2010. Volcanic ash as fertiliser for the surface ocean. Atmospheric Chemistry and Physics 10:3891-3899; doi:10.5194/acp-10-3891-2010.

Lara, L.E. 2009. The 2008 eruption of the Chaitén Volcano, Chile: A preliminary report. Andean Geology 36(1):125-129.

Laughlin, A.W., J. Poths, H.A. Healey, S. Reneau, and G. WoldeGabriel. 1994. Dating of Quaternary basalts using the cosmogenic ^{3}He and ^{14}C methods with implications for excess ^{40}Ar. Geology 22(2):135-138; doi:10.1130/0091-7613(1994)022<0135:DOQBUT>2.3.CO;2.

Lavallée, Y., P.M. Benson, M.J. Heap, K.U. Hess, A. Flaws, B. Schillinger, P.G. Meredith, and D.B. Dingwell. 2013. Reconstructing magma failure and the degassing network of dome-building eruptions. Geology 41(4):515-518; doi:0.1130/G33948.1.

LeGrande, A., K. Tsigaridis, and S.E. Bauer. 2016. Role of atmospheric chemistry in the climate impacts of stratospheric volcanic injections. Nature Geoscience 9:652-655; doi:10.1038/ngeo2771.

Lin, G., P.M. Shearer, F. Amelung, and P.G. Okubo. 2015. Seismic tomography of compressional wave attenuation structure for Kīlauea Volcano, Hawai'i. Journal of Geophysical Research: Solid Earth 120:2510-2524; doi:10.1002/2014JB011594.

Lin, I.I., C. Hu, Y.H. Li, T.Y. Ho, T.P. Fischer, G.T.F. Wong, J. Wu, C.W. Huang, D.A. Chu, D.S. Ko, and J.P. Chen. 2011. Fertilization potential of volcanic dust in the low-nutrient low-chlorophyll western North Pacific subtropical gyre: Satellite evidence and laboratory study. Global Biogeochemical Cycles 25(1):GB1006; doi:10.1029/2009GB003758.

Linde, A.T., and I.S. Sacks. 1998. Triggering of volcanic eruptions. Nature 395(6705):888-890; doi:10.1038/27650.

Linde, A.T., K. Agustsson, S. Sacks, and R. Stefansson. 1993. Mechanism of the 1991 eruption of Hekla from continuous borehole strain monitoring. Nature 365(6448):737-740; doi:10.1038/365737a0.

Lindoo, A., J.F. Larsen, K.V. Cashman, A.L. Dunn, and O.K. Neill. 2016. An experimental study of permeability development as a function of crystal-free melt viscosity. Earth and Planetary Science Letters 435:45-54; doi:10.1016/j.epsl.2015.11.035.

Lipman, P.W., and N.G. Banks. 1987. Aa flow dynamics, Mauna Loa 1984. Pp. 1527-1567 in Volcanism in Hawaii, Vol. 2, R.W. Decker, T.L. Wright, and P.H. Stauffer, eds. U.S. Geological Survey Professional Paper 1350. Available at https://pubs.usgs.gov/pp/1987/1350/pdf/chapters/pp1350_ch57.pdf. Accessed December 13, 2016.

Lipman, P.W., and D.R. Mullineaux, eds. 1981. The 1980 Eruptions of Mount St. Helens, Washington. U.S. Geological Survey Professional Paper 1250. Available at https://pubs.usgs.gov/pp/1250/report.pdf. Accessed December 13, 2016.

Liu, E.J., K.V. Cashman, F.M. Beckett, C.S. Witham, S.J. Leadbetter, M.C. Hort, and S. Guðmundsson. 2014. Ash mists and brown snow: Remobilization of volcanic ash from recent Icelandic eruptions. Journal of Geophysical Research: Atmospheres 119(15):9463-9480; doi:10.1002/2014JD021598.

Liu, E.J., K.V. Cashman, A.C. Rust, and S.R. Gislason. 2015. The role of bubbles in generating fine ash during hydromagmatic eruptions. Geology 43(3):239-242; doi:10.1130/G36336.1.

Lloyd, A.S., P. Ruprecht, E.H. Hauri, W. Rose, H.M. Gonnermann, and T. Plank. 2014. NanoSIMS results from olivine-hosted melt embayments: Magma ascent rate during explosive basaltic eruptions. Journal of Volcanology and Geothermal Research 283:1-18; doi:10.1016/j.jvolgeores.2014.06.002.

Long, C.J., M.J. Power, T.A. Minckley, and A.L. Hass. 2014. The impact of Mt Mazama tephra deposition on forest vegetation in the Central Cascades, Oregon, USA. The Holocene 24(4):503-511; doi:10.1177/0959683613520258.

Loughlin, S.C., R. Luckett, G. Ryan, T. Christopher, V. Hards, S. De Angelis, L. Jones, and M. Strutt. 2010. An overview of lava dome evolution, dome collapse and cyclicity at Soufrière Hills Volcano, Montserrat, 2005-2007. Geophysical Research Letters 37(19):L00E16; doi:10.1029/2010GL042547.

Loughlin, S.C., C. Vye-Brown, R.S.J. Sparks, S.K. Brown, and S. Jenkins, eds. 2015. Global Volcanic Hazards and Risk. Cambridge, UK: Cambridge University Press. Available at http://globalvolcanomodel.org/wp-content/uploads/2015/08/Global-Volcanic-Hazards-and-Risk-Full-book-low-res.pdf. Accessed March 1, 2017.

Lowenstern, J., W.C. Evans, D. Bergfeld, and A.G. Hunt. 2014. Prodigious degassing of a billion years of accumulated radiogenic helium at Yellowstone. Nature 506(7488):355-358; doi:10.1038/nature12992.

Lu, Z., and D. Dzurisin. 2014. InSAR Imaging of Aleutian Volcanoes: Monitoring a Volcanic Arc from Space. Heidelberg: Springer. doi:10.1007/978-3-642-00248-6.

Mackinnon, I.D.R., J.L. Gooding, D.S. McKay, and U.S. Clanton. 1984. The El Chichon stratospheric cloud — Solid particulates and settling rates. Journal of Volcanology and Geothermal Research 23(1-2):125-146; doi:10.1016/0377-0273(84)90059-3.

Maclennan, J., M. Jull, D.P. McKenzie, L. Slater, and K. Gronvold. 2002. The link between volcanism and deglaciation in Iceland. Geochemistry, Geophysics, Geosystems 3(11):1062; doi:10.1029/2001GC000282.

Mader, H.M., M. Manga, and T. Koyaguchi. 2004. The role of laboratory experiments in volcanology. Journal of Volcanology and Geothermal Research 129(1):1-5; doi:10.1016/S0377-0273(03)00228-2.

Maeda, I. 2000. Nonlinear visco-elastic volcanic model and its application to the recent eruption of Mt. Unzen. Journal of Volcanology and Geothermal Research 95(1-4):35-47; doi:10.1016/S0377-0273(99)00120-1.

Maeno, F., S. Nakada, and T. Kaneko. 2016. Morphological evolution of a new volcanic islet sustained by compound lava flows. Geology 44(4):259-262; doi:10.1130/G37461.1.

Major, J.J., T.C. Pierson, R.L. Dinehart, and J.E. Costa. 2000. Sediment yield following severe volcanic disturbance: A two-decade perspective from Mount St Helens. Geology 28(9):819-822; doi:10.1130/0091-7613(2000)28<819:SYFSVD>2.0.CO;2.

Malone, S.D., E.T. Endo, C.S. Weaver, and J.W. Ramey. 1981. Seismic monitoring for eruption prediction. Pp. 803-813 in The 1980 Eruptions of Mount St. Helens, Washington, P.W. Lipman and D.R. Mullineux, eds. U.S. Geological Survey Professional Paper 1250. Available at https://pubs.usgs.gov/pp/1250/report.pdf. Accessed December 13, 2016.

Manga, M., and E. Brodsky. 2006. Seismic triggering of eruptions in the far field: Volcanoes and geysers. Annual Review of Earth and Planetary Sciences 34:263-291; doi:10.1146/annurev.earth.34. 031405.125125.

Mangan, M., and T. Sisson. 2000. Delayed, disequilibrium degassing in rhyolite magma: Decompression experiments and implications for explosive volcanism. Earth and Planetary Science Letters 183(3-4):441-455; doi:10.1016/S0012-821X(00)00299-5.

Mannen, K. 2014. Particle segregation of an eruption plume as revealed by a comprehensive analysis of tephra dispersal: Theory and application. Journal of Volcanology and Geothermal Research 284:61-78; doi:10.1016/j.jvolgeores.2014.07.009.

Manville, V., and S.J. Cronin. 2007. Breakout lahar from New Zealand's crater lake. EOS 88(43):441-442; doi:10.1029/2007EO430001.

Manzella, I., C. Bonadonna, J.C. Phillips, and H. Monnard. 2015. The role of gravitational instabilities in deposition of volcanic ash. Geology 43(3):211-214; doi:10.1130/G36252.1.

Marsh, B.D. 2007. Crystallization of silicate magmas deciphered using crystal size distributions. Journal of the American Ceramic Society 90(3):746-757; doi:10.1111/j.1551-2916.2006.01473.x.

Marsh, B., W. Teplow, M. Reagan, and K. Sims. 2008. Puna dacite: Likely temperature, viscosity, origin, size and parent body nature. Abstract V32A2130. American Geophysical Union Fall Meeting.

Marzocchi, W., and M.S. Bebbington. 2012. Probabilistic eruption forecasting at short and long time scales. Bulletin of Volcanology 74(8):1777-1805; doi:10.1007/s00445-012-0633-x.

Marzocchi, W., L. Sandri, and J. Selva. 2008. BET_EF: A probabilistic tool for long- and short-term eruption forecasting. Bulletin of Volcanology 70(5):623-632; doi:10.1007/s00445-007-0157-y.

Mason, B.G., D.M. Pyle, W.B. Dade, and T. Jupp. 2004. Seasonality of volcanic eruptions. Journal of Geophysical Research: Solid Earth 109(B4):B04206; doi:10.1029/2002JB002293.

Mastin, L.G., M. Guffanti, R. Servranckx, P. Webley, S. Barsotti, K. Dean, A. Durant, J.W. Ewert, A. Neri, W.I. Rose, D. Schneider, L. Siebert, B. Stunder, G. Swanson, A. Tupper, A. Volentik, and C.F. Waythomas. 2009a. A multidisciplinary effort to assign realistic source parameters to models of volcanic ash-cloud transport and dispersion during eruptions. Journal of Volcanology and Geothermal Research 186(1):10-21; doi:10.1016/j.jvolgeores.2009.01.008.

Mastin, L.G., O. Spieler, and W.S. Downey. 2009b. An experimental study of hydromagmatic fragmentation through energetic, non-explosive magma–water mixing. Journal of Volcanology and Geothermal Research 180(2-4):161-170; doi:10.1016/j.jvolgeores.2008.09.012.

Mastin, L.G., A.R. Van Eaton, and J.B. Lowenstern. 2014. Modeling ash fall distribution from a Yellowstone supereruption. Geochemistry, Geophysics, Geosystems 15(8):3459-3475; doi:10.1002/2014GC005469.

Mather, T.A. 2015. Volcanoes and the environment: Lessons for understanding Earth's past and future from studies of present-day volcanic emissions. Journal of Volcanology and Geothermal Research 304:160-179; doi:10.1016/j.jvolgeores.2015.08.016.

Matthews, A.J., J. Barclay, S. Carn, G. Thompson, J. Alexander, R. Herd, and C. Williams. 2002. Rainfall-induced volcanic activity on Montserrat. Geophysical Research Letters 29(13):22-1–22-4; doi:10.1029/2002GL014863.

Matthews, A.J., J. Barclay, and J.E. Johnstone. 2009. The fast response of volcano-seismic activity to intense precipitation: Triggering of primary volcanic activity by rainfall at Soufrière Hills Volcano, Montserrat. Journal of Volcanology and Geothermal Research 184(3-4):405-415; doi:10.1016/j.jvolgeores.2009.05.010.

Mattox, T.N., C. Heliker, J. Kauahikaua, and K. Hon. 1993. Development of the 1990 Kalapana flow field, Kilauea Volcano, Hawaii. Bulletin of Volcanology 55(6):407-413; doi:10.1007/BF00302000.

McCormick, M.P., L.W. Thomason, and C.R. Trepte. 1995. Atmospheric effects of the Mt Pinatubo eruption. Nature 373(6513):399-404; doi:10.1038/373399a0.

McGuire, W.J., R.J. Howarth, C.R. Firth, A.R. Solow, A.D. Pullen, S.J. Saunders, I.S. Steward, and C. Vita-Finzi. 1997. Correlation between rate of sea level change and frequency of explosive volcanism in the Mediterranean. Nature 389(6650):473-476; doi:10.1038/38998.

McNutt, S.R., and E.R. Williams. 2010. Volcanic lightning: Global observations and constraints on source mechanisms. Bulletin of Volcanology 72(10):1153-1167; doi:10.1007/s00445-010-0393-4.

Melnik, O., and R.S.J. Sparks. 2002. Dynamics of magma ascent and lava extrusion at Soufrière Hills Volcano, Montserrat. Memoirs 21(1):153-171; doi:10.1144/GSL.MEM.2002.021.01.07.

Melnik, O., and R.S.J. Sparks. 2006. Controls on conduit magma flow dynamics during lava dome building eruptions. Journal of Geophysical Research: Solid Earth 110(B2):B02209; doi:10.1029/2004JB003183.

Merucci, L., K. Zakšek, E. Carboni, and S. Corradini. 2016. Stereoscopic estimation of volcanic ash cloud-top height from two geostationary satellites. Remote Sensing 8(3):206; doi:10.3390/rs8030206.

Michaut, C., Y. Ricard, D. Bercovici, and R.S.J. Sparks. 2013. Eruption cyclicity at silicic volcanoes potentially caused by magmatic gas waves. Nature Geoscience 6(10):856-860; doi:10.1038/ngeo1928.

Michieli Vitturi, D.M., A.B. Clarke, A. Neri, and B. Voight. 2008. Effects of conduit geometry on magma ascent dynamics in dome-forming eruptions. Earth and Planetary Science Letters 272(3-4):567-578; doi:10.1016/j.epsl.2008.05.025.

Miller, A.D., R.C. Stewart, R.A. White, R. Luckett, B.J. Baptie, W.P. Aspinall, J.L. Latchman, L.L. Lynch, and B. Voight. 1998. Seismicity associated with dome growth and collapse at the Soufriere Hills Volcano, Montserrat. Geophysical Research Letters 25(18):3401-3404; doi:10.1029/98GL01778.

Miller, C.F., D.J. Furbish, B.A. Walker, L.L. Claiborne, G.C. Koteas, H.A. Bleick, and J.S. Miller. 2011. Growth of plutons by incremental emplacement of sheets in crystal-rich host: Evidence from Miocene intrusions of the Colorado River region, Nevada, USA. Tectonophysics 500(1-4):65-77; doi:10.1016/j.tecto.2009.07.011.

Moitra, P., and H.M. Gonnermann. 2015. Effects of crystal shape- and size-modality on magma rheology. Geochemistry, Geophysics, Geosystems 16(1):1-26; doi:10.1002/2014GC005554.

Moore, J.G. 1967. Base surge in recent volcanic eruptions. Bulletin Volcanologique 30(1):337-363; doi:10.1007/BF02597678.

Moran, S.C., C. Newhall, and D.C. Roman. 2011. Failed magmatic eruptions: Late-stage cessation of magma ascent. Bulletin of Volcanology 73(2):115-122; doi:10.1007/s00445-010-0444-x.

Morgan, D.J., S. Blake, N.W. Rogers, B. De Vivo, G. Rolandi, and J.P. Davidson. 2006. Magma chamber recharge at Vesuvius in the century prior to the eruption of A.D. 79. Geology 34(10):845-848; doi:10.1130/G22604.1.

Morrissey, M., B. Zimanowski, K. Wohletz, and R. Buettner. 2000. Phreatomagmatic fragmentation. Pp. 431-445 in Encyclopedia of Volcanoes, H. Sigurdsson, B. Houghton, S.R. McNutt, H. Rymer, and J. Stix, eds. San Diego, CA: Academic Press. Available at http://www.geo.auth.gr/yliko/useful/books/books_geology/E/Encyclopedia%20of%20Volcanoes.pdf. Accessed December 13, 2016.

Mortensen, A.K., Þ. Egilson, B. Bautason, S. Árnadóttir, and Á. Guðmundsson. 2014. Stratigraphy, alteration mineralogy, permeability and temperature conditions of well IDDP-1, Krafla, NE-Iceland. Geothermics 49:31-41; doi:10.1016/j.geothermics.2013.09.013.

Mueller, S., E.W. Llewellin, and H.M. Mader. 2011. The effect of particle shape on suspension viscosity and implications for magmatic flows. Geophysical Research Letters 38(13):L13316; doi:10.1029/2011GL047167.

Muller, J.R., G. Ito, and S.J. Martel. 2001. Effects of volcano loading on dike propagation in an elastic half-space. Journal of Geophysical Research: Solid Earth 106(B6):11101-11113; doi:10.1029/2000JB900461.

Murtagh, R.M., and J.D. White. 2013. Pyroclast characteristics of a subaqueous to emergent Surtseyan eruption, Black Point volcano, California. Journal of Volcanology and Geothermal Research 267:75-91; doi:10.1016/j.jvolgeores.2013.08.015.

Nadeau, P., J. Palma, and G. Waite. 2011. Linking volcanic tremor, degassing, and eruption dynamics via SO_2 imaging. Geophysical Research Letters 38(1):L01304; doi:10.1029/2010GL045820.

Nakada, S., H. Shimizu, and K. Ohta. 1999. Overview of the 1990-1995 eruption at Unzen Volcano. Journal of Volcanology and Geothermal Research 89(1-4):1-22; doi:10.1016/S0377-0273(98)00118-8.

Nakada, S., K. Uto, S. Sakuma, J.C. Eichelberger, and H. Shimizu. 2005. Scientific results of conduit drilling in the Unzen scientific drilling project (USDP). Scientific Drilling 1:18-22; doi:10.2204/iodp.sd.1.03.2005.

Namiki, A., and M. Manga. 2008. Transition between fragmentation and permeable outgassing of low viscosity magmas. Journal of Volcanology and Geothermal Research 169(1-2):48-60; doi:10.1016/j.jvolgeores.2007.07.020.

Namiki, A., E. Rivalta, H. Woith, and T.R. Walter. 2016. Sloshing of a bubbly magma reservoir as a mechanism of triggered eruptions. Journal of Volcanology and Geothermal Research 320:156-171; doi:10.1016/j.jvolgeores.2016.03.010.

Naranjo, J.A., R.S.J. Sparks, M.V. Stasiuk, H. Moreno, and G.J. Ablay. 1992. Morphological, structural and textural variations in the 1988–1990 andesite lava of Lonquimay Volcano, Chile. Geological Magazine 129(6):657-678; doi:10.1017/S0016756800008426.

Neave, D.A., J. Maclennan, T. Thordarson, and M.E. Hartley. 2015. The evolution and storage of primitive melts in the Eastern Volcanic Zone of Iceland: The 10 ka Grímsvötn tephra series (i.e. the Saksunarvatn ash). Contributions to Mineralogy and Petrology 170(2):21; doi:10.1007/s00410-015-1170-3.

Neri, A., and F. Dobran. 1994. Influence of eruption parameters on the thermofluid dynamics of collapsing volcanic columns. Journal of Geophysical Research: Solid Earth 99(B6):11833-11857; doi:10.1029/94JB00471.

Neri, A., and G. Macedonio. 1996. Numerical simulation of collapsing volcanic columns with particles of two sizes. Journal of Geophysical Research: Solid Earth 101(B4):8153-8174; doi:10.1029/95JB03451.

Neri, A., W.P. Aspinall, R. Cioni, A. Bertagnini, P.J. Baxter, G. Zuccaro, D. Andronico, S. Barsotti, P.D. Cole, T.E. Ongaro, T.K. Hincks, G. Macedonio, P. Papale, M. Rosi, R. Santacroce, and G. Woo. 2008. Developing an event tree for probabilistic hazard and risk assessment at Vesuvius. Journal of Volcanology and Geothermal Research 178(3):397-415; doi:10.1016/j.jvolgeores.2008.05.014.

Neri, A., A. Bevilacqua, T.E. Ongaro, R. Isaia, W.P. Aspinall, M. Bisson, F. Flandoli, P.J. Baxter, A. Bertagnini, E. Iannuzzi, S. Orsucci, M. Pistolesi, M. Rosi, and S. Vitale. 2015. Quantifying volcanic hazard at Campi Flegrei caldera (Italy) with uncertainty assessment: 2. Pyroclastic density current invasion maps. Journal of Geophysical Research: Solid Earth 120(4):2330-2349; doi:10.1002/2014JB011776.

Newhall, C.G., and S. Self. 1982. The Volcanic Explosivity Index (VEI): An estimate of explosive magnitude for historical volcanism. Journal of Geophysical Research: Oceans 87(C2):1231-1238; doi:10.1029/JC087iC02p01231.

Newhall, C.G., A.S. Daag, F.G. Delfin, R.P. Hoblitt, J. McGeehin, J.S. Pallister, T.M. Regalado, M. Rubin, B.S. Tubianosa, R.A. Tamayo, and J.V. Umbal. 1996. Eruptive history of Mount Pinatubo. Pp. 165-195 in Fire and Mud: Eruptions and Lahars of Mount Pinatubo, Philippines. Quezon City: Philippine Institute of Volcanology and Seismology. Available at https://pubs.usgs.gov/pinatubo/contents.html. Accessed December 14, 2016.

Nishimura, T., M. Iguchi, R. Kawaguchi, M. Hendrasto, and U. Rosadi. 2012. Inflations prior to Vulcanian eruptions and gas bursts detected by tilt observations at Semeru Volcano, Indonesia. Bulletin of Volcanology 74(4):903-911; doi:10.1007/s00445-012-0579-z.

Nostro, C., R.S. Stein, M. Cocco, M.E. Belardinelli, and W. Marzocchi. 1998. Two-way coupling between Vesuvius eruptions and southern Apennine earthquakes, Italy, by elastic stress transfer. Journal of Geophysical Research: Solid Earth 103(B10):24487-24504; doi:10.1029/98JB00902.

Oberhuber, J.M., M. Herzog, H.F. Graf, and K. Schwanke. 1998. Volcanic plume simulation on large scales. Journal of Volcanology and Geothermal Research 87(1-4):29-53; doi:10.1016/S0377-0273(98)00099-7.

Obermann, A., T. Planès, E. Larose, and M. Campillo. 2013. Imaging preeruptive and coeruptive structural and mechanical changes of a volcano with ambient seismic noise. Journal of Geophysical Research: Solid Earth 118(12):6285-6294; doi:10.1002/2013JB010399.

Ogburn, S.E., J. Berger, E.S. Calder, D. Lopes, A. Patra, E.B. Pitman, R. Rutarindwa, E. Spiller, and R.L. Wolpert. 2016. Pooling strength amongst limited datasets using hierarchical Bayesian analysis, with application to pyroclastic density current mobility metrics. Statistics in Volcanology 2(1):1-26; doi:10.5038/2163-338X.2.1.

Okubo, P.G., and C.J. Wolfe. 2008. Swarms of similar long-period earthquakes in the mantle beneath Mauna Loa Volcano. Journal of Volcanology and Geothermal Research 178(4):787-794; doi:10.1016/j.jvolgeores.2008.09.007.

Okumura, S., M. Nakamura, S. Takeuchi, A. Tsuchiyama, T. Nakano, and K. Uesugi. 2009. Magma deformation may induce non-explosive volcanism via degassing through bubble networks. Earth and Planetary Science Letters 281(3-4):267-274; doi:10.1016/j.epsl.2009.02.036.

Oman, L., A. Robock, G.L. Stenchikov, and T. Thordarson. 2006. High-latitude eruptions cast shadow over the African monsoon and the flow of the Nile. Geophysical Research Letters 33(18):L18711; doi:10.1029/2006GL027665.

Ongaro, T.E., C. Widiwijayanti, A.B. Clarke, B. Voight, and A. Neri. 2011. Multiphase-flow numerical modeling of the 18 May 1980 lateral blast at Mount St. Helens, USA. Geology 39(6):535-538; doi:10.1130/G31865.1.

Oppenheimer, C. 2002. Limited global change due to the largest known Quaternary eruption, Toba ≈74 kyr BP? Quaternary Science Reviews 21(14-15):1593-1609; doi:10.1016/S0277-3791(01)00154-8.

Oppenheimer, J., A.C. Rust, K.V. Cashman, and B. Sandnes. 2015. Gas migration regimes and outgassing in particle-rich suspensions. Frontiers in Physics 3:60; doi:10.3389/fphy.2015.00060.

Ozerov, A., I. Ispolatov, and J. Lees. 2003. Modeling Strombolian eruptions of Karymsky volcano, Kamchatka, Russia. Journal of Volcanology and Geothermal Research 122(3-4):265-280; doi:10.1016/S0377-0273(02)00506-1.

Pallister, J.S., K.V. Cashman, J.T. Hagstrum, N.M. Beeler, S.C. Moran, and R.P. Denlinger. 2013. Faulting within the Mount St. Helens conduit and implications for volcanic earthquakes. Geological Society of America Bulletin 125(3-4):359-376; doi:10.1130/B30716.1.

Paris, R. 2015. Source mechanisms of volcanic tsunamis. Philosophical Transactions of the Royal Society A 373(2053):20140380; doi:10.1098/rsta.2014.0380.

Passarelli, L., and E.E. Brodsky. 2012. The correlation between run-up and repose times of volcanic eruptions. Geophysical Journal International 188(3):1025-1045; doi:10.1111/j.1365-246X.2011.05298.x.

Patanè, D., A. Aiuppa, M. Aloisi, B. Behncke, A. Cannata, M. Coltelli, G. Di Grazia, S. Gambino, S. Gurrieri, M. Mattia, and G. Salerno. 2013. Insights into magma and fluid transfer at Mount Etna by a multiparametric approach: A model of the events leading to the 2011 eruptive cycle. Journal of Geophysical Research: Solid Earth 118(7):3519-3539; doi:10.1002/jgrb.50248.

Patrick, M.R., T. Orr, L. Antolik, L. Lee, and K. Kamibayashi. 2014. Continuous monitoring of Hawaiian volcanoes with thermal cameras. Journal of Applied Volcanology 3(1):1; doi:10.1186/2191-5040-3-1.

Pavolonis, M.J., A.K. Heidinger, and J. Sieglaff. 2013. Automated retrievals of volcanic ash and dust cloud properties from upwelling infrared measurements. Journal of Geophysical Research: Atmospheres 118(3):1436-1458; doi:10.1002/jgrd.50173.

Peacock, J.R., M.T. Mangan, D. McPhee, and P.E. Wannamaker. 2016. Three-dimensional electrical resistivity model of the hydrothermal system in Long Valley Caldera, California, from magnetotellurics. Geophysical Research Letters 43(15):7953-7962; doi:10.1002/2016GL069263.

Peate, D.W., and C.J. Hawkesworth. 2005. U series disequilibria: Insights into mantle melting and the timescales of magma differentiation. Reviews of Geophysics 43(1):RG1003; doi:10.1029/2004RG000154.

Peltier, A., V. Ferrazzini, T. Staudacher, and P. Bachèlery. 2005. Imaging the dynamics of dyke propagation prior to the 2000–2003 flank eruptions at Piton de La Fournaise, Reunion Island. Geophysical Research Letters 32(22):L22302; doi:10.1029/2005GL023720.

Phillipson, G., R. Sobradelo, and J. Gottsmann. 2013. Global volcanic unrest in the 21st century: An analysis of the first decade. Journal of Volcanology and Geothermal Research 264:183-196; doi:10.1016/j.jvolgeores.2013.08.004.

Pierson, T., and J. Major. 2014. Hydrogeomorphic effects of explosive eruptions on drainage basins. Annual Review of Earth and Planetary Science 42:469-507; doi:10.1146/annurev-earth-060313-054913.

Pierson, T.C., R.J. Janda, J.C. Thouret, and C.A. Borrero. 1990. Perturbation and melting of snow and ice by the 13 November 1985 eruption of Nevado del Ruiz, Colombia, and consequent mobilization, flow and deposition of lahars. Journal of Volcanology and Geothermal Research 41(1-4):17-66; doi:10.1016/0377-0273(90)90082-Q.

Pinel, V., and C. Jaupart. 2003. Magma chamber behavior beneath a volcanic edifice. Journal of Geophysical Research: Solid Earth 108(B2):2072; doi:10.1029/2002JB001751.

Pinel, V., C. Jaupart, and F. Albino. 2010. On the relationship between cycles of eruptive activity and growth of a volcanic edifice. Journal of Volcanology and Geothermal Research 194(4):150-164; doi:10.1016/j.jvolgeores.2010.05.006.

Pioli, L., B.J. Azzopardi, and K.V. Cashman. 2009. Controls on the explosivity of scoria cone eruptions: Magma segregation at conduit junctions. Journal of Volcanology and Geothermal Research 186(3-4):407-415; doi:10.1016/j.jvolgeores.2009.07.014.

Pioli, L., C. Bonadonna, B.J. Azzopardi, J.C. Phillips, and M. Ripepe. 2012. Experimental constraints on the outgassing dynamics of basaltic magmas. Journal of Geophysical Research: Solid Earth 117(B3):B03204; doi:10.1029/2011JB008392.

Pistone, M., L. Caricchi, P. Ulmer, L. Burlini, P. Ardia, E. Reusser, F. Marone, and L. Arbaret. 2012. Deformation experiments of bubble-and crystal-bearing magmas: Rheological and microstructural analysis. Journal of Geophysical Research: Solid Earth 117(B5):B05208; doi:10.1029/2011JB008986.

Plag, H.P., S. Blocklebank, D. Brosnan, P. Campus, S. Cloetingh, S. Jules-Plag, and S. Stein. 2015. Extreme Geohazards: Reducing the Disaster Risk and Increasing Resilience. European Science Foundation. Available at http://www.esf.org/fileadmin/Public_documents/Publications/Natural_Hazards.pdf. Accessed December 14, 2016.

Plank, T., K.A. Kelley, M.M. Zimmer, E.H. Hauri, and P.J. Wallace. 2013. Why do mafic arc magmas contain ~4 wt% water on average? Earth and Planetary Science Letters 364:168-179; doi:10.1016/j.epsl.2012.11.044.

Poland, M.P., A. Miklius, and E.K. Montgomery-Brown. 2014. Magma supply, storage, and transport at shield-stage Hawaiian volcanoes. Pp. 179-234 in Characteristics of Hawaiian Volcanoes, M.P. Poland, T.J. Takahashi, and C.M. Landowski, eds. U.S. Geological Survey Professional Paper 1801. Available at https://pubs.usgs.gov/pp/1801/downloads/pp1801_Chap5_Poland.pdf. Accessed December 14, 2016.

Pollock, N.M., B.D. Brand, and O. Roche. 2016. The controls and consequences of substrate entrainment by pyroclastic density currents at Mount St Helens, Washington (USA). Journal of Volcanology and Geothermal Research 325:135-147; doi:10.1016/j.jvolgeores.2016.06.012.

Power, J.A., S.D. Stihler, R.A. White, and S.C. Moran. 2004. Observations of deep long-period (DLP) seismic events beneath Aleutian arc volcanoes; 1989-2002. Journal of Volcanology and Geothermal Research 138(3-4):243-266; doi:10.1016/j.jvolgeores.2004.07.005.

Power, J.A., M.L. Coombs, and J.T. Freymueller, eds. 2010. The 2006 Eruption of Augustine Volcano, Alaska. U.S. Geological Survey Professional Paper 1769. Available at https://pubs.usgs.gov/pp/1769. Accessed December 14, 2016.

Pritchard, M.E., and P.M. Gregg. 2016. Geophysical evidence for silicic crustal melt in the continents: Where, what kind, and how much? Elements 12(2):121-127; doi:10.2113/gselements.12.2.121.

Pritchard, M.E., J.A. Jay, F. Aron, S.T. Henderson, and L.E. Lara. 2013. Subsidence at southern Andes volcanoes induced by the 2010 Maule, Chile earthquake. Nature Geoscience 6(8):632-636; doi:10.1038/ngeo1855.

Proussevitch, A.A., and D.L. Sahagian. 1998. Dynamics and energetics of bubble growth in magmas: Analytical formulation and numerical modeling. Journal of Geophysical Research: Solid Earth 103(B8):18223-18251; doi:10.1029/98JB00906.

Punongbayan, R.S., C.G. Newhall, M.L.P. Bautista, D. Garcia, D.H. Harlow, R.P. Hoblitt, J.P. Sabit, and R.U. Solidum. 1996. Eruption hazard assessments and warnings. Pp. 67-85 in Fire and Mud: Eruptions and Lahars of Mount Pinatubo, Philippines. Quezon City: Philippine Institute of Volcanology and Seismology. Available at https://pubs.usgs.gov/pinatubo/contents.html. Accessed December 14, 2016.

Putirka, K., and F.J. Tepley III. 2008. Minerals, Inclusions and Volcanic Processes. Reviews in Mineralogy and Geochemistry Vol. 69. Chantilly, VA: Mineralogical Society of America.

Pyle, D.M. 2015. Sizes of volcanic eruptions. Pp. 257-264 in Encyclopedia of Volcanoes, 2nd Ed., H. Sigurdsson, B. Houghton, S.R. McNutt, H. Rymer, and J. Stix, eds. San Diego, CA: Academic Press. doi:10.1016/B978-0-12-385938-9.00013-4.

Rae, A.S.P., M. Edmonds, J. Maclennan, D. Morgan, B. Houghton, M.E. Hartley, and I. Sides. 2016. Time scales of magma transport and mixing at Kīlauea Volcano, Hawai'i. Geology 44(6):463-466; doi:10.1130/g37800.1.

Rampino, M.R., and S. Self. 2015. Large igneous provinces and biotic extinctions. Pp. 1049-1058 in Encyclopedia of Volcanoes, 2nd Ed., H. Sigurdsson, B. Houghton, S.R. McNutt, H. Rymer, and J. Stix, eds. San Diego, CA: Academic Press. doi:10.1016/B978-0-12-385938-9.00061-4.

Rawson, H., D.M. Pyle, T.A. Mather, V.C. Smith, K. Fontijn, S.M. Lachowycz, and J.A. Naranjo. 2016. The magmatic and eruptive response of arc volcanoes to deglaciation: Insights from southern Chile. Geology 44(4):251-254; doi:10.1130/G37504.1.

Reath, K.A., M.S. Ramsey, J. Dehn, and P.W. Webley. 2016. Predicting eruptions from precursory activity using remote sensing data hybridization. Journal Volcanology and Geothermal Research 321:18-30; doi:10.1016/j.jvolgeores.2016.04.027.

Reid, M.R. 2003. Timescales of magma transfer and storage in the crust. Pp. 167-193 in The Crust, R.L. Rudnick, ed. Treatise on Geochemistry Vol. 3. Amsterdam: Elsevier. doi:10.1016/B0-08-043751-6/03022-X.

Reid, M.R., and J.A. Vazquez. 2017. Fitful and protracted magma assembly leading to a giant eruption, Youngest Toba Tuff, Indonesia. Geochemistry, Geophysics, Geosystems 18(1):156-177; doi:10.1002/2016GC006641.

Resing, J.A., K.H. Rubin, R. Embley, J. Lupton, E. Baker, R. Dziak, T. Baumberger, M. Lilley, J. Huber, T.M. Shank, D. Butterfield, D. Clague, N. Keller, S. Merle, N.J. Buck, P. Michael, A. Soule, D. Caress, S. Walker, R. Davis, J. Cowen, A.L. Reysenbach, and H. Thomas. 2011. Active submarine eruption of boninite in the northeastern Lau Basin. Nature Geosciences 4:799-806; doi:10.1038/ngeo1275.

Ripepe, M., D. Delle Donne, R. Genco, G. Maggio, M. Pistolesi, E. Marchetti, G. Lacanna, G. Ulivieri, and P. Poggi. 2015. Volcano seismicity and ground deformation unveil the gravity-driven magma discharge dynamics of a volcanic eruption. Nature Communications 6:6998; doi:10.1038/ncomms7998.

Rivalta, E., B. Taisne, A.P. Bunger, and R.F. Katz. 2015. A review of mechanical models of dike propagation: Schools of thought, results and future directions. Tectonophysics 638:1-42; doi:10.1016/j.tecto.2014.10.003.

Robertson, E.A.M., J. Biggs, K.V. Cashman, M.A. Floyd, and C. Vye-Brown. 2016. Influence of regional tectonics and pre-existing structures on the formation of elliptical calderas in the Kenyan Rift. Pp. 43-68 in Magmatic Rifting and Active Volcanism, T.J. Wright, A. Ayele, D.J. Ferguson, T. Kidane, and C. Vye-Brown, eds. Special Publications 420. London: Geological Society. doi:10.1144/SP420.12.

Robertson, R.E.A., P. Cole, R.S.J. Sparks, C. Harford, A.M. Lejeune, W.J. McGuire, A.D. Miller, M.D. Murphy, G. Norton, N.F. Stevens, and S.R. Young. 1998. The explosive eruption of Soufriere Hills Volcano, Montserrat, West Indies, 17 September, 1996. Geophysical Research Letters 25(18):3429-3432; doi:10.1029/98GL01442.

Robock, A. 2000. Volcanic eruptions and climate. Review of Geophysics 38(2):191-219; doi:10.1029/1998RG000054.

Robock, A., L. Oman, and G.L. Stenchikov. 2008. Regional climate responses to geoengineering with tropical and Arctic SO_2 injections. Journal of Geophysical Research: Atmospheres 113(D16):D16101; doi:10.1029/2008JD010050.

Robock, A., A. Marquardt, B. Kravitz, and G. Stenchikov. 2009. Benefits, risks, and costs of stratospheric geoengineering. Geophysical Research Letters 36(19):L19703; doi:10.1029/2009GL039209.

Roche, O., D.C. Buesch, and G.A. Valentine. 2016. Slow-moving and far-travelled dense pyroclastic flows during the Peach Spring super-eruption. Nature Communications 7:10890; doi:10.1038/ncomms10890.

Rodgers, M., D.C. Roman, H. Geirsson, P. LaFemina, S.R. McNutt, A. Muñoz, and V. Tenorio. 2015. Stable and unstable phases of elevated seismic activity at the persistently restless Telica Volcano, Nicaragua. Journal of Volcanology and Geothermal Research 290:63-74; doi:10.1016/j.jvolgeores.2014.11.012.

Rodolfo, K.S., J.V. Umbal, R.A. Alonso, C.T. Remotigue, M.L. Paladio-Melosantos, J.H.G. Salvador, D. Evangelista, and Y. Miller. 1996. Two years of lahars on the western flank of Mount Pinatubo: Initiation, flow processes, deposits, and attendant geomorphic and hydraulic changes. Pp. 989-1013 in Fire and Mud: Eruptions and Lahars of Mount Pinatubo, Philippines. Quezon City: Philippine Institute of Volcanology and Seismology. Available at https://pubs.usgs.gov/pinatubo/contents.html. Accessed December 15, 2016.

Roman, D.C., and K.V. Cashman. 2006. The origin of volcano-tectonic earthquake swarms. Geology 34(6):457-460; doi:10.1130/G22269.1.

Romero, J.E., D. Morgavi, F. Arzilli, R. Daga, A. Caselli, F. Reckziegel, J. Viramonte, J. Diaz-Alvarado, M. Polacci, M. Burton, and D. Perugini. 2016. Eruption dynamics of the 22-23 April 2015 Calbuco Volcano (Southern Chile): Analyses of tephra fall deposits. Journal of Volcanology and Geothermal Research 317:15-29; doi:10.1016/j.jvolgeores.2016.02.027.

Rose, W.I., and A.J. Durant. 2011. Fate of volcanic ash: Aggregation and fallout. Geology 39(9):895-896; doi:10.1130/focus092011.1.

Rosen, J. 2016. Crystal clocks. Science 354(6314):822-825; doi:10.1126/science.354.6314.822.

Rougier, J., S.R. Sparks, and K.V. Cashman. 2016. Global recording rates for large eruptions. Journal of Applied Volcanology 5(1):11; doi:10.1186/s13617-016-0051-4.

Rubin, A.M. 1993. On the thermal viability of dikes leaving magma chambers. Geophysical Research Letters 20(4):257-260; doi:10.1029/92GL02783.

Rubin, A.M. 1995. Propagation of magma-filled cracks. Annual Review of Earth and Planetary Sciences 23:287-336; doi:10.1146/annurev.ea.23.050195.001443.

Rubin, A.M., D. Gillard, and J.L. Got. 1998. A reinterpretation of seismicity associated with the January 1983 dike intrusion at Kilauea Volcano, Hawaii. Journal of Geophysical Research: Solid Earth 103(B5):10003-10015; doi:10.1029/97JB03513.

Rubin, K.H., S.A. Soule, W.W. Chadwick, Jr., D.J. Fornari, D.A. Clague, R.W. Embley, E.T. Baker, M.R. Perfit, D.W. Caress, and R.P. Dziak. 2012. Volcanic eruptions in the deep sea. Oceanography 25(1):142-157; doi:10.5670/oceanog.2012.12.

Rust, A.C., and K.V. Cashman. 2004. Permeability of vesicular silicic magma: Inertial and hysteresis effects. Earth and Planetary Science Letters 228(1-2):93-107; doi:10.1016/j.epsl.2004.09.025.

Rust, A.C., and K.V. Cashman. 2011. Permeability controls on expansion and size distributions of pyroclasts. Journal of Geophysical Research: Solid Earth 116(B11):B11202; doi:10.1029/2011JB008494.

Rutherford, M.J. 2008. Magma ascent rates. Reviews in Mineralogy and Geochemistry 69(1):241-271; doi:10.2138/rmg.2008.69.7.

Sable, J.E., B.F. Houghton, P. Del Carlo, and M. Coltelli. 2006. Changing conditions of magma ascent and fragmentation during the Etna 122 BC basaltic Plinian eruption: Evidence from clast microtextures. Journal of Volcanology and Geothermal Research 158(3-4):333-354; doi:10.1016/j.jvolgeores.2006.07.006.

Sable, J.E., B.F. Houghton, C.J.N. Wilson, and R.J. Carey. 2009. Eruption mechanisms during the climax of the Tarawera 1886 basaltic Plinian eruption inferred from microtextural characteristics of the deposits. Pp. 129-154 in Studies in Volcanology: The Legacy of George Walker, T. Thordarson, S. Self, G. Larsen, S.K. Rowland, and A. Hoskuldsson, eds. Special Publication of IAVCEI 2. London: Geological Society. doi:10.1144/IAVCEI002.

Sahagian, D.L. 2005. Volcanic eruption mechanisms: Insights from intercomparison of models of conduit processes. Journal of Volcanology and Geothermal Research 143(1-3):1-15; doi:10.1016/j.jvolgeores.2004.12.006.

Sakuma, S., T. Kajiwara, S. Nakada, K. Uto, and H. Shimizu. 2008. Drilling and logging results of USDP-4: Penetration into the volcanic conduit of Unzen Volcano, Japan. Journal of Volcanology and Geothermal Research 175(1-2):1-12; doi:10.1016/j.jvolgeores.2008.03.039.

Saltzman, M.R., and E. Thomas. 2012. Carbon isotope stratigraphy. Pp. 207-232 in The Geologic Time Scale 2012, M.F. Gradstein, J.G. Ogg, M.D. Schmitz, and G.M. Ogg, eds. Amsterdam: Elsevier. doi:10.1016/B978-0-444-59425-9.00011-1.

Santer, B.D., C. Bonfils, J.P. Painter, M.D. Zelinka, C. Mears, S. Solomon, G.A. Schmidt, J.C. Fyfe, J.N.S. Cole, L. Nazarenko, K.E. Taylor, and F.J. Wentz. 2014. Volcanic contribution to decadal changes in tropospheric temperature. Nature Geoscience 7:185-189; doi:10.1038/ngeo2098.

Saunders, K., J. Blundy, R. Dohmen, and K. Cashman. 2012. Linking petrology and seismology at an active volcano. Science 336(6084):1023-1027; doi:10.1126/science.1220066.

Scandone, R., K.V. Cashman, and S.D. Malone. 2007. Magma supply, magma ascent and the style of volcanic eruptions. Earth and Planetary Science Letters 253(3-4):513-529; doi:10.1016/j.epsl.2006.11.016.

Scharff, L., M. Hort, and N.R. Varley. 2015. Pulsed Vulcanian explosions: A characterization of eruption dynamics using Doppler radar. Geology 43(11):995-998; doi:10.1130/G36705.1.

Schmidt, A., K.S. Carslaw, G.W. Mann, M. Wilson, T.J. Breider, S.J. Pickering, and T. Thodarson. 2010. The impact of the 1783-1784 AD Laki eruption on global aerosol formation processes and cloud condensation nuclei. Atmosheric Chemistry and Physics 10:6025-6041; doi:10.5194/acp-10-6025-2010.

Schmidt, A., R.A. Skeffington, T. Thordarson, S. Self, P.M. Forster, A. Rap, A. Ridgwell, D. Fowler, M. Wilson, G.M. Mann, P.B. Wignall, and K.S. Carslaw. 2016. Selective environmental stress from sulphur emitted by continental flood basalt eruptions. Nature Geoscience 9:77-82; doi:10.1038/ngeo2588.

Schmitt, A.K. 2011. Uranium series accessory crystal dating of magmatic processes. Annual Review of Earth and Planetary Sciences 39:321-349; doi:10.1146/annurev-earth-040610-133330.

Schneider, D.J., and R.P. Hoblitt. 2013. Doppler weather radar observations of the 2009 eruption of Redoubt Volcano, Alaska. Journal of Volcanology and Geothermal Research 259:133-144; doi:10.1016/j.jvolgeores.2012.11.004.

Schwaiger, H.F., R.P. Denlinger, and L.G. Mastin. 2012. Ash3d: A finite-volume, conservative numerical model for ash transport and tephra deposition. Journal of Geophysical Research: Solid Earth 117(B4):B04204; doi:10.1029/2011JB008968.

Scollo, S., A. Folch, and A. Costa. 2008a. A parametric and comparative study of different tephra fallout models. Journal of Volcanology and Geothermal Research 176(2):199-211; doi:10.1016/j.jvolgeores.2008.04.002.

Scollo, S., S. Tarantola, C. Bonadonna, M. Coltelli, and A. Saltelli. 2008b. Sensitivity analysis and uncertainty estimation for tephra dispersal models. Journal of Geophysical Research: Solid Earth 113(B6):B06202; doi:10.1029/2006JB004864.

Scott, W.E., D.R. Sherrod, and C.A. Gardner. 2008. Overview of the 2004 to 2006, and continuing, eruption of Mount St. Helens, Washington. Pp. 3-22 in A Volcano Rekindled: The Renewed Eruption of Mount St. Helens, 2004-2006, D.R. Sherrod, W.E. Scott, and P.H. Stauffer, eds. U.S. Geological Survey Professional Paper 1750. Available at https://pubs.usgs.gov/pp/1750/chapters/pp2008-1750_chapter01.pdf. Accessed December 15, 2016.

Segall, P. 2013. Volcano deformation and eruption forecasting. Pp. 85-106 in Remote Sensing of Volcanoes and Volcanic Processes: Integrating Observation and Modelling, D.M. Pyle, T.A. Mather, and J. Biggs, eds. Special Publications 380. London: Geological Society. doi:10.1144/SP380.4.

Segall, P., A.L. Llenos, S.H. Yun, A.M. Bradley, and E.M. Syracuse. 2013. Time-dependent dike propagation from joint inversion of seismicity and deformation data. Journal of Geophysical Research: Solid Earth 118(11):5785-5804; doi:10.1002/2013JB010251.

Sehlke, A., A. Whittington, B. Robert, A. Harris, L. Gurioli, and E. Médard. 2014. Pahoehoe to 'a'a transition of Hawaiian lavas: An experimental study. Bulletin of Volcanology 76:876; doi:10.1007/s00445-014-0876-9.

Self, S. 2006. The effects and consequences of very large explosive volcanic eruptions. Philosophical Transactions of the Royal Society A 364(1845):2073-2097; doi:10.1098/rsta.2006.1814.

Self, S., M.R. Rampino, M.S. Newton, and J.A. Wolff. 1984. Volcanological study of the great Tambora eruption of 1815. Geology 12(11):659-663; doi:10.1130/0091-7613(1984)12<659:VSOTGT> 2.0.CO;2.

Self, S., T. Thordarson, and L. Kelzthelyi. 1997. Emplacement of continental flood basalt lava flows. Pp. 381-410 in Large Igneous Provinces: Continental, Oceanic, and Planetary Flood Volcanism, M.F. Coffin and J.J. Mahoney, eds. Geophysical Monograph 100. Washington, DC: American Geophysical Union. doi:10.1029/GM100p0381.

Self, S., A. Schmidt, and T.A. Mather. 2014. Emplacement characteristics, timescales, and volatile release rates of continental flood basalt eruptions on Earth. Pp. 319-337 in Volcanism, Impacts, and Mass Extinctions: Causes and Effects, G. Keller and A.C. Kerr, eds. Geological Society of America Special Paper 505. doi:10.1130/2014.2505(16).

Sheldrake, T.E., R.S.J. Sparks, K.V. Cashman, G. Wadge, and W.P. Aspinall. 2016. Similarities and differences in the historical records of lava dome-building volcanoes: Implications for understanding magmatic processes and eruption forecasting. Earth-Science Reviews 160:240-263; doi:10.1016/j.earscirev.2016.07.013.

Sheridan, M.F., and K.H. Wohletz. 1983. Hydrovolcanism: Basic considerations and review. Journal of Volcanology and Geothermal Research 17(1-4):1-29; doi:10.1016/0377-0273(83)90060-4.

Sherrod, D.R., W.E. Scott, and P.H. Stauffer, eds. 2008. A Volcano Rekindled: The Renewed Eruption of Mount St. Helens, 2004-2006. U.S. Geological Survey Professional Paper 1750. Available at https://pubs.usgs.gov/pp/1750. Accessed December 16, 2016.

Shinohara, H. 2008. Excess degassing from volcanoes and its role on eruptive and intrusive activity. Reviews of Geophysics 46(4):RG4005; doi:10.1029/2007RG000244.

Sides, I., M. Edmonds, J. Maclennan, B.F. Houghton, D.A. Swanson, and M.J. Steele-MacInnis. 2014. Magma mixing and high fountaining during the 1959 Kilauea Iki eruption, Hawai'i. Earth and Planetary Science Letters 400:102-112; doi:10.1016/j.epsl.2014.05.024.

Siebert, L., E. Cottrell, E. Venzke, and B. Andrews. 2015. Earth's volcanoes and their eruptions: An overview. Pp. 239-255 in Encyclopedia of Volcanoes, 2nd Ed., H. Sigurdsson, B. Houghton, S.R. McNutt, H. Rymer, and J. Stix, eds. San Diego, CA: Academic Press. doi:10.1016/B978-0-12-385938-9.00012-2.

Sigmarsson, O., B. Haddadi, S. Carn, S. Moune, J. Gudnason, K. Yang, and L. Clarisse. 2013. The sulfur budget of the 2011 Grímsvötn eruption, Iceland. Geophysical Research Letters 40(23):6095-6100; doi:10.1002/2013GL057760.

Sigmundsson, F., A. Hooper, S. Hreinsdóttir, and 34 others. 2015. Segmented lateral dyke growth in a rifting event at Bárðarbunga volcanic system, Iceland. Nature 517(7533):191-195; doi:10.1038/nature14111.

Sigurdsson, H., B. Houghton, S.R. McNutt, H. Rymer, and J. Stix, eds. 2015. Encyclopedia of Volcanoes, 2nd Ed. San Diego, CA: Academic Press.

Sillitoe, R.H. 2010. Porphyry copper systems. Economic Geology 105(1):3-41; doi:0.2113/gsecongeo.105.1.3.

Simmons, J., D. Elsworth, and B. Voight. 2005. Classification and idealized limit-equilibrium analyses of dome collapses at Soufrière Hills volcano, Montserrat, during growth of the first lava dome: November 1995-March 1998. Journal of Volcanology and Geothermal Research 139(3-4):241-258; doi:10.1016/j.jvolgeores.2004.08.009.

Solomon, S., J.S. Daniel, R.R. Neely III, J.P. Vernier, E.G. Dutton, and L.W. Thomason. 2011. The persistently variable background stratospheric aerosol layer and global climate change. Science 333(6044):866-870; doi:10.1126/science.1206027.

Solovitz, S.A., D.E. Ogden, D.D.W. Kim, and S.Y. Kim. 2014. Coupled fluid and solid evolution in analogue volcanic vents. Journal of Geophysical Research: Solid Earth 119(7):5342-5355; doi:10.1002/2014JB010993.

Soule, S.A., D.J. Fornari, M.R. Perfit, and K.H. Rubin. 2007. New insights into mid-ocean ridge volcanic processes from the 2005–2006 eruption of the East Pacific Rise, 9°46′N–9°56′N. Geology 35(12):1079-1082; doi:10.1130/G23924A.1.

Sparks, R.S.J. 1978. The dynamics of bubble formation and growth in magmas: A review and analysis. Journal of Volcanology and Geothermal Research 3(1-2):1-37; doi:10.1016/0377-0273(78)90002-1.

Sparks, R.S.J. 1986. The dimensions and dynamics of volcanic eruption columns. Bulletin of Volcanology 48(1):3-15; doi:10.1007/BF01073509.

Sparks, R.S.J. 2003. Forecasting volcanic eruptions. Earth and Planetary Science Letters 210(1-2):1-15; doi:10.1016/S0012-821X(03)00124-9.

Sparks, R.S.J., and S.R. Young. 2002. The eruption of Soufriere Hills Volcano, Montserrat (1995-1999): Overview of scientific results. Memoirs 21(1):45-69; doi:10.1144/GSL.MEM.2002.021.01.03.

Sparks, R.S.J., S.C. Loughlin, E. Cottrell, G. Valentine, C. Newhall, G. Jolly, P. Papale, S. Takarada, S. Crosweller, M. Nayembil, and B. Arora. 2012. Global Volcano Model. P. 13299 in European Geoscience Union General Assembly Conference, April 22-27, 2012, Vienna, Austria, Research Abstracts, Vol. 14. Available at http://meetingorganizer.copernicus.org/EGU2012/EGU2012-13299.pdf. Accessed December 16, 2016.

Spieler, O., B. Kennedy, U. Kueppers, D.B. Dingwell, B. Scheu, and J. Taddeucci. 2004. The fragmentation threshold of pyroclastic rocks. Earth and Planetary Science Letters 226(1-2):139-148; doi:10.1016/j.epsl.2004.07.016.

Stefánsson, R. 2011. Advances in Earthquake Prediction, Research and Risk Mitigation. Berlin: Springer-PRAXIS, 271 pp.

Stix, J., M. Calvache, T.P. Fischer, D. Gómez, L. Narvaez, M. Ordoñez, E.A. Ortega, C.R. Torres, and S.N. Williams. 1993. A model of degassing at Galeras Volcano, Colombia, 1988-1993. Geology 21(11):963-967; doi:10.1130/0091-7613(1993)021<0963:AMODAG>2.3.CO;2.

Stohl, A., A.J. Prata, S. Eckhardt, L. Clarisse, A. Durant, S. Henne, N.I. Kristiansen, A. Minikin, U. Schumann, P. Seibert, K. Stebel, H.E. Thomas, T. Thorsteinsson, K. Tørseth, and B. Weinzierl. 2011. Determination of time- and height-resolved volcanic ash emissions and their use for quantitative ash dispersion modeling: The 2010 Eyjafjallajökull eruption. Atmospheric Chemistry and Physics 11:4333-4351; doi:10.5194/acp-11-4333-2011.

Stroberg, T.W., M. Manga, and J. Dufek. 2010. Heat transfer coefficients of natural volcanic clasts. Journal of Volcanology and Geothermal Research 194:214-219; doi:10.1016/j.jvolgeores.2010.05.007.

Sumita, I., and M. Manga. 2008. Suspension rheology under oscillatory shear and its geophysical implications. Earth and Planetary Science Letters 269(3):468-477; doi:10.1016/j.epsl.2008.02.043.

Suzuki, Y.J., and T. Koyaguchi. 2013. 3D numerical simulation of volcanic eruption clouds during the 2011 Shinmoe-dake eruptions. Earth, Planets and Space 65(6):581-589; doi:10.5047/eps.2013.03.009.

Svensen, H., S. Planke, L. Chevallier, A. Malthe-Sørenssen, F. Corfu, and B. Jamtveit. 2007. Hydrothermal venting of greenhouse gases triggering early Jurassic global warming. Earth and Planetary Science Letters 256(3-4):554-566; doi:10.1016/j.epsl.2007.02.013.

Svensen, H., S. Planke, A.G. Polozov, N. Schmidbauer, F. Corfu, Y.Y. Podladchikov, and B. Jamtveit. 2009. Siberian gas venting and the end-Permian environmental crisis. Earth and Planetary Science Letters 277(3-4):490-500; doi:10.1016/j.epsl.2008.11.015.

Swanson, D.A., T.J. Casadevall, D. Dzurisin, S.D. Malone, C.G. Newhall, and C.S. Weaver. 1983. Predicting eruptions at Mount St. Helens, June 1980 through December 1982. Science 221(4618):1369-1376; doi:10.1126/science.221.4618.1369.

Swanson, D.A., T.R. Rose, A.E. Mucek, M.O. Garcia, R.S. Fiske, and L.G. Mastin. 2014. Cycles of explosive and effusive eruptions at Kīlauea Volcano, Hawai'i. Geology 42(7):631-634; doi:10.1130/G35701.1.

Syracuse, E.M., M. Maceira, H. Zhang, and C.H. Thurber. 2015. Seismicity and structure of Akutan and Makushin Volcanoes, Alaska, using joint body and surface wave tomography. Journal of Geophysical Research: Solid Earth 120(2):1036-1052; doi:10.1002/2014JB011616.

Taddeucci, J., P. Scarlato, A. Capponi, E. Del Bello, C. Cimarelli, D.M. Palladino, and U. Kueppers. 2012. High-speed imaging of Strombolian explosions: The ejection velocity of pyroclasts. Geophysical Research Letters 39(2):L02302; doi:10.1029/2011GL050404.

Taisne, B., F. Brenguier, N.M. Shapiro, and V. Ferrazzini. 2011. Imaging the dynamics of magma propagation using radiated seismic intensity. Geophysical Research Letters 38(4):L04304; doi:10.1029/2010GL046068.

Tait, S., C. Jaupart, and S. Vergniolle. 1989. Pressure, gas content and eruption periodicity of a shallow, crystallising magma chamber. Earth and Planetary Science Letters 92(1):107-123; doi:10.1016/0012-821X(89)90025-3.

Takada, Y., and Y. Fukushima. 2013. Volcanic subsidence triggered by the 2011 Tohoku earthquake in Japan. Nature Geoscience 6(8):637-641; doi:10.1038/ngeo1857.

Takeuchi, S., S. Nakashima, and A. Tomiya. 2008. Permeability measurements of natural and experimental volcanic materials with a simple permeameter: Toward an understanding of magmatic degassing processes. Journal of Volcanology and Geothermal Research 177(2):329-339; doi:10.1016/j.jvolgeores.2008.05.010.

Tarasewicz, J., R.S. White, A.W. Woods, B. Brandsdóttir, and M.T. Gudmundsson. 2012. Magma mobilization by downward-propagating decompression of the Eyjafjallajökull volcanic plumbing system. Geophysical Research Letters 39(19):L19309; doi:10.1029/2012GL053518.

Taron, J., D. Elsworth, G. Thompson, and B. Voight. 2007. Mechanisms for rainfall-concurrent lava dome collapses at Soufrière Hills Volcano, 2000-2002. Journal of Volcanology and Geothermal Research 160(1-2):195-209; doi:10.1016/j.jvolgeores.2006.10.003.

Tarquini, S., and M. de'Michieli Vitturi. 2014. Influence of fluctuating supply on the emplacement dynamics of channelized lava flows. Bulletin of Volcanology 76(3):801; doi:10.1007/s00445-014-0801-2.

Thordarson, T., and S. Self. 2003. Atmospheric and environmental effects of the 1783-1784 Laki eruption: A review and reassessment. Journal of Geophysical Research 108(D1):4011; doi:10.1029/2001JD002042.

Timmreck, C. 2012. Modeling the climatic effects of large explosive volcanic eruptions. Wiley Interdisciplinary Reviews: Climate Change 3(6):545-564; doi:10.1002/wcc.192.

Timmreck, C., S.J. Lorenz, T.J. Crowley, S. Kinne, T.J. Raddatz, M.A. Thomas, and J.H. Jungclaus. 2009. Limited temperature response to the very large AD 1258 volcanic eruption. Geophysical Research Letters 36(21):L21708; doi:10.1029/2009GL040083.

Torres, R.C., S. Self, and M.M.L. Martinez. 1996. Secondary pyroclastic flows from the June 15, 1991, ignimbrite of Mount Pinatubo. Pp. 625-678 in Fire and Mud: Eruptions and Lahars of Mount Pinatubo, Philippines. Quezon City: Philippine Institute of Volcanology and Seismology. Available at https://pubs.usgs.gov/pinatubo/contents.html. Accessed December 15, 2016.

Trenberth, K.E., and A. Dai. 2007. Effects of Mount Pinatubo volcanic eruption on the hydrological cycle as an analog of geoengineering. Geophysical Research Letters 34(15):L15702; doi:10.1029/2007GL030524.

Tuffen, H. 2010. How will melting of ice affect volcanic hazards in the twenty-first century? Philosophical Transactions of the Royal Society A 368(1919):2535-2558; doi:10.1098/rsta.2010.0063.

Tuffen, H., D.B. Dingwell, and H. Pinkerton. 2003. Repeated fracture and healing of silicic magma generate flow banding and earthquakes? Geology 31(12):1089-1092; doi:10.1130/G19777.1.

Tuffen, H., M.R. James, J.M. Castro, and C.I. Schipper. 2013. Exceptional mobility of an advancing rhyolitic obsidian flow at Cordón Caulle volcano in Chile. Nature Communications 4:2709; doi:10.1038/ncomms3709.

Turner, S., and F. Costa. 2007. Measuring timescales of magmatic evolution. Elements 3(4):267-272; doi:10.2113/gselements.3.4.267.

Ulberg, C.W., K.C. Creager, S.C. Moran, G.A. Abers, R.P. Denlinger, A.J. Hotovec-Ellis, J.E. Vidale, E. Kiser, A. Levander, and A. Schultz. 2014. Imaging Magma Under St. Helens (iMUSH): Details of passive-source seismic deployment and preliminary 3-D velocity structure. Abstract S11E-4387. American Geophysical Union Fall Meeting.

Valentine, G.A. 1987. Stratified flow in pyroclastic surges. Bulletin of Volcanology 49(4):616-630; doi:10.1007/BF01079967.

Valentine, G.A. 1998. Eruption column physics. Pp. 91-138 in From Magma to Tephra: Modelling Physical Processes of Explosive Volcanic Eruptions, A. Freundt and M. Rosi, eds. Amsterdam: Elsevier.

Valentine, G.A., and J.D. White. 2012. Revised conceptual model for maar-diatremes: Subsurface processes, energetics, and eruptive products. Geology 40(12):1111-1114; doi:10.1130/G33411.1.

Valentine, G.A., and K.H. Wohletz. 1989. Numerical models of Plinian eruption columns and pyroclastic flows. Journal of Geophysical Research: Solid Earth 94(B2):1867-1887; doi:10.1029/JB094iB02p01867.

Valentine, G.A., C. Bonadonna, I. Manzella, A. Clarke, and P. Dellino. 2011. Large-scale experiments on volcanic processes. EOS 92(11):89-90; doi:10.1029/2011EO110001.

Vallance, J.W., and R.M. Iverson. 2015. Lahars and their deposits. Pp. 649-664 in Encyclopedia of Volcanoes, 2nd Ed., H. Sigurdsson, B. Houghton, S.R. McNutt, H. Rymer, and J. Stix, eds. San Diego, CA: Academic Press. doi:10.1016/B978-0-12-385938-9.00037-7.

Van Daele, M., J. Moernaut, G. Silversmit, S. Schmidt, K. Fontijn, K. Heirman, W. Vandoorne, M. De Clercq, J. Van Acker, C. Wolff, M. Pino, R. Urrutia, S.J. Roberts, L. Vincze, and M. De Batist. 2014. The 600 yr eruptive history of Villarrica Volcano (Chile) revealed by annually laminated lake sediments. Geological Society of America Bulletin 126(3-4):481-498; doi:10.1130/B30798.1.

Van Eaton, A.R., L.G. Mastin, M. Herzog, H.F. Schwaiger, D.J. Schneider, K.L. Wallace, and A.B. Clarke. 2015. Hail formation triggers rapid ash aggregation in volcanic plumes. Nature Communications 6:7860; doi:10.1038/ncomms8860.

Van Eaton, A.R., Á. Amigo, D. Bertin, L.G. Mastin, R.E. Giacosa, J. González, O. Valderrama, K. Fontikn, and S.A. Behnke. 2016. Volcanic lightning and plume behavior reveal evolving hazards during the April 2015 eruption of Calbuco volcano, Chile. Geophysical Research Letters 43(7):3563-3571; doi:10.1002/2016GL068076.

Venzke, E., ed. 2013. Volcanoes of the World, v. 4.5.2. Smithsonian Institution. doi:10.5479/si.GVP.VOTW4-2013.

Vernier, J.P., L.W. Thomason, J.P. Pommereau, A. Bourassa, J. Pelon, A. Garnier, A. Hauchecorne, L. Blanot, C. Trepte, D. Degenstein, and F. Vargas. 2011. Major influence of tropical volcanic eruptions on the stratospheric aerosol layer during the last decade. Geophysical Research Letters 38(12):L12807; doi:10.1029/2011GL047563.

Vernier, J.P., T.D. Fairlie, T. Deshler, M. Natarajan, T. Knepp, K. Foster, F.G. Wienhold, K.M. Bedka, L. Thomason, and C. Trepte. 2016. In situ and space-based observations of the Kelud volcanic plume: The persistence of ash in the lower stratosphere. Journal of Geophysical Research: Atmospheres 121(18):11104-11118: doi:10.1002/2016JD025344.

Vinkler, A.P., K. Cashman, G. Giordano, and G. Groppelli. 2012. Evolution of the mafic Villa Senni caldera-forming eruption at Colli Albani volcano, Italy, indicated by textural analysis of juvenile fragments. Journal of Volcanology and Geothermal Research 235:37-54; doi:10.1016/j.jvolgeores.2012.03.006.

Violette, S., G. De Marsily, J.P. Carbonnel, P. Goblet, E. Ledoux, S.M. Tijani, and G. Vouille. 2001. Can rainfall trigger volcanic eruptions? A mechanical stress model of an active volcano: "Piton de la Fournaise," Reunion Island. Terra Nova 13(1):18-24; doi:10.1046/j.1365-3121.2001.00297.x.

Voight, B. 1988. A method for prediction of volcanic eruptions. Nature 332(6160):125-130; doi:10.1038/332125a0.

Voight, B. 1990. The 1985 Nevado del Ruiz volcano catastrophe: Anatomy and retrospection. Journal of Volcanology and Geothermal Research 44(3-4):349-386; doi:10.1016/0377-0273(90)90027-D.

Voight, B., and D. Elsworth. 2000. Instability and collapse of hazardous gas-pressurized lava domes. Geophysical Research Letters 27(1):1-4; doi:10.1029/1999GL008389.

Voight, B., R.P. Hoblitt, A.B. Clarke, A.B. Lockhart, A.D. Miller, L. Lynch, and J. McMahon. 1998. Remarkable cyclic ground deformation monitored in real time on Montserrat and its use in eruption forecasting. Geophysical Research Letters 25(18):3405-3408; doi:10.1029/98GL01160.

Wadge, G., P.A.V. Young, and I.J. McKendrick. 1994. Mapping lava flow hazards using computer simulation. Journal of Geophysical Research: Solid Earth 99(B1):489-504; doi:10.1029/93JB01561.

Wadge, G., P.W. Francis, and C.F. Ramirez. 1995. The Socompa collapse and avalanche event. Journal of Volcanology and Geothermal Research 66(1):309-336; doi:10.1016/0377-0273(94)00083-S.

Wadge, G., R. Herd, G. Ryan, E.S. Calder, and J.C. Komorowski. 2010. Lava production at Soufrière Hills Volcano, Montserrat: 1995-2009. Geophysical Research Letters 37(19):L00E03; doi:10.1029/2009GL041466.

Wadge, G., B. Voight, R.S.J. Sparks, P.D. Cole, S.C. Loughlin, and R.E.A. Robertson. 2014. An overview of the eruption of Soufriere Hills Volcano, Montserrat from 2000 to 2010. Memoirs 39(1):1-40; doi:10.1144/M39.1.

Walker, G.P.L. 1971. Compound and simple lava flows and flood basalts. Bulletin Volcanologique 35(3):579-590; doi:10.1007/BF02596829.

Walker, G.P.L. 1973. Explosive volcanic eruptions—A new classification scheme. Geologische Rundschau 62(2):431-446; doi:10.1007/ BF01840108.

Walker, L.R., D.S. Sikes, A.R. Degange, S.C. Jewett, G. Michaelson, S.L. Talbot, S.S. Talbot, B. Wang, and J.C. Williams. 2013. Biological legacies: Direct early ecosystem recovery and food web reorganization after a volcanic eruption in Alaska. Ecoscience 20(3):240-251; doi:10.2980/20-3-3603.

Wallace, P.J. 2005. Volatiles in subduction zone magmas: Concentrations and fluxes based on melt inclusion and volcanic gas data. Journal of Volcanology and Geothermal Research 140(1):217-240; doi:10.1016/j.jvolgeores.2004.07.023.

Walter, T.R., and F. Amelung. 2007. Volcanic eruptions following M≥ 9 megathrust earthquakes: Implications for the Sumatra-Andaman volcanoes. Geology 35(6):539-542; doi:10.1130/G23429A.1.

Walter, T.R., R. Wang, M. Zimmer, H. Grosser, B. Lühr, and A. Ratdomopurbo. 2007. Volcanic activity influenced by tectonic earthquakes: Static and dynamic stress triggering at Mt. Merapi. Geophysical Research Letters 34(5):L05304; doi:10.1029/2006GL028710.

Watkins, J.M., M. Manga, and D.J. DePaolo. 2012. Bubble geobarometry: A record of pressure changes, degassing, and regassing at Mono Craters, California. Geology 40(8):699-702; doi:10.1130/G33027.1.

Watt, S.F.L., D.M. Pyle, and T.A. Mather. 2009. The influence of great earthquakes on volcanic eruption rate along the Chilean subduction zone. Earth and Planetary Science Letters 277(3-4):399-407; doi:10.1016/j.epsl.2008.11.005.

Watt, S.F.L., D.M. Pyle, and T.A. Mather. 2013. The volcanic response to deglaciation: Evidence from glaciated arcs and a reassessment of global eruption records. Earth-Science Reviews 122:77-102; doi:10.1016/j.earscirev.2013.03.007.

Watt, S.F., J.S. Gilbert, A. Folch, J.C. Phillips, and X.M. Cai. 2015. An example of enhanced tephra deposition driven by topographically induced atmospheric turbulence. Bulletin of Volcanology 77:35; doi:10.1007/s00445-015-0927-x.

Watts, R.B., R.A. Herd, R.S.J. Sparks, and S.R. Young. 2002. Growth patterns and emplacement of the andesitic lava dome at Soufriere Hills Volcano, Montserrat. Memoirs 21(1):115-152; doi:10.1144/GSL.MEM.2002.021.01.06.

Waythomas, C.F., T.P. Miller, and M.T. Mangan. 2006. Preliminary Volcano Hazard Assessment for the Emmons Lake Volcanic Center, Alaska. U.S. Geological Survey Scientific Investigations Report 2006–5248. Available at https://pubs.usgs.gov/sir/2006/5248/pdf/SIR2006-5248.pdf. Accessed April 13, 2017.

West, M., J.J. Sánchez, and S.R. McNutt. 2005. Periodically triggered seismicity at Mount Wrangell, Alaska, after the Sumatra earthquake. Science 308(5725):1144-1146; doi:10.1126/science.1112462.

White, J.D.L., and B. Houghton. 2000. Surtseyan and related phreatomagmatic eruptions. Pp. 495-512 in Encyclopedia of Volcanoes, H. Sigurdsson, B. Houghton, S.R. McNutt, H. Rymer, and J. Stix, eds. San Diego, CA: Academic Press. Available at http://www.geo.auth.gr/yliko/useful/books/books_geology/E/Encyclopedia%20of%20Volcanoes.pdf. Accessed December 13, 2016.

White, J.D., and P.S. Ross. 2011. Maar-diatreme volcanoes: A review. Journal of Volcanology and Geothermal Research 201(1-4):1-29; doi:10.1016/j.jvolgeores.2011.01.010.

White, R., and W. McCausland. 2016. Volcano-tectonic earthquakes: A new tool for estimating intrusive volumes and forecasting eruptions. Journal of Volcanology and Geothermal Research 309:139-155; doi:10.1016/j.jvolgeores.2015.10.020.

White, R.S., J. Drew, H.R. Martens, J. Key, H. Soosalu, and S.S. Jakobsdóttir. 2011. Dynamics of dyke intrusion in the mid-crust of Iceland. Earth and Planetary Science Letters 304(3):300-312; doi:10.1016/j.epsl.2011.02.038.

White, S.M., J.A. Crisp, and F.J. Spera. 2006. Long-term volumetric eruption rates and magma budgets. Geochemistry, Geophysics, Geosystems 7(3); doi:10.1029/2005GC001002.

Wilcock, W.S., M. Tolstoy, F. Waldhauser, C. Garcia, Y.J. Tan, D.R. Bohnenstiehl, J. Caplan-Auerbach, R.P. Dziak, A.F. Arnulf, and M.E. Mann. 2016. Seismic constraints on caldera dynamics from the 2015 Axial Seamount eruption. Science 354(6318):1395-1399; doi:10.1126/science.aah5563.

Wilson, C.J.N. 1980. The role of fluidization in the emplacement of pyroclastic claws: An experimental approach. Journal of Volcanology and Geothermal Research 8(2-4):231-249; doi:10.1016/0377-0273(80)90106-7.

Wilson, L., R.S.J. Sparks, T.C. Huang, and N.D. Watkins. 1978. The control of eruption column heights by eruption energetics and dynamics. Journal of Geophysical Research: Solid Earth 83(B4):1829-1836; doi:10.1029/JB083iB04p01829.

Wilson, L., R.S.J. Sparks, and G.P.L. Walker. 1980. Explosive volcanic eruptions- IV. The control magma properties and conduit geometry on eruption column behaviour. Geophysical Journal International 63(1):117-148; doi:10.1111/j.1365-246X.1980.tb02613.x.

Wilson, T.M., J.W. Cole, C. Stewart, S.J. Cronin, and D.M. Johnston. 2011. Ash storms: Impacts of wind-remobilised volcanic ash on rural communities and agriculture following the 1991 Hudson eruption, southern Patagonia, Chile. Bulletin of Volcanology 73(3):223-239; doi:10.1007/s00445-010-0396-1.

Winson, A.E., F. Costa, C.G. Newhall, and G. Woo. 2014. An analysis of the issuance of volcanic alert levels during volcanic crises. Journal of Applied Volcanology 3:14; doi:10.1186/s13617-014-0014-6.

Wohletz, K., B. Zimanowski, and R. Büttner. 2012. Magma–water interactions. P. 230 in Modeling Volcanic Processes: The Physics and Mathematics of Volcanism, S.A. Fagents, T.K.P. Gregg, and R.M.C Lopes, eds. New York: Cambridge University Press. Available at http://www.lanl.gov/orgs/ees/geodynamics/Wohletz/2013_Magma-WaterInteractions.pdf. Accessed December 15, 2016.

Woo, G. 2008. Probabilistic criteria for volcano evacuation decisions. Natural Hazards 45(1):87-97; doi:10.1007/s11069-007-9171-9.

Woods, A.W. 1988. The fluid dynamics and thermodynamics of eruption columns. Bulletin of Volcanology 50(3):169-193; doi:10.1007/BF01079681.

Wright, R., L.P. Flynn, H. Garbeil, A.J. Harris, and E. Pilger. 2004. MODVOLC: Near-real-time thermal monitoring of global volcanism. Journal of Volcanology and Geothermal Research 135(1-2):29-49; doi:10.1016/j.jvolgeores.2003.12.008.

Wright, T.J., C. Ebinger, J. Biggs, A. Ayele, G. Yirgu, D. Keir, and A. Stork. 2006. Magma-maintained rift segmentation at continental rupture in the 2005 Afar dyking episode. Nature 442(7100):291-294; doi:10.1038/nature04978.

Zablocki, C.J., R.I. Tilling, D.W. Peterson, R.L. Christiansen, G.V. Keller, and J.C. Murray. 1974. A deep research drill hole at the summit of an active volcano, Kilauea, Hawaii. Geophysical Research Letters 1(7):323-326; doi:10.1029/GL001i007p00323.

Zellmer, G.F., S. Blake, D. Vance, C. Hawkesworth, and S. Turner. 1999. Plagioclase residence times at two island arc volcanoes (Kameni Islands, Santorini, and Soufriere, St. Vincent) determined by Sr diffusion systematics. Contributions to Mineralogy and Petrology 136(4):345-357; doi:10.1007/s004100050543.

Zellmer, G.F., C. Annen, B.L.A. Charlier, R.M.M. George, S.P. Turner, and C.J. Hawkesworth. 2005. Magma evolution and ascent at volcanic arcs: Constraining petrogenetic processes through rates and chronologies. Journal of Volcanology and Geothermal Research 140(1-3):171-191; doi:10.1016/j.jvolgeores.2004.07.020.

Zhang, Y. 1996. Dynamics of CO_2-driven lake eruptions. Nature 379(6560):57-59; doi:10.1038/379057a0.

Zhang, Y. 1999. H_2O in rhyolitic glasses and melts: Measurement, speciation, solubility, and diffusion. Reviews of Geophysics 37(4):493-516; doi:10.1029/1999RG900012.

Zierenberg, R.A., P. Schiffman, G.H. Barfod, C.E. Lesher, N.E. Marks, J.B. Lowenstern, A.K. Mortensen, E.C. Pope, G.Ó. Friðleifsson, and W.A. Elders. 2012. Origin and emplacement of a rhyolite melt into the Krafla Geothermal System, Iceland. Contributions to Mineralogy and Petrology 165(2):327-347; doi:10.1007/s00410-012-0811-z.

Zimanowski, B., and R. Buttner. 2003. Phreatomagmatic explosions in subaqueous volcanism. Pp. 51-60 in Submarine Explosive Volcanism, J.D.L. White, J.L. Smellie, and D.A. Clague, eds. Geophysical Monograph 140. Washington, DC: American Geophysical Union. Available at https://www.researchgate.net/profile/James_White12/publication/228074178_Introduction_A_deductive_outline_and_topical_overview_of_subaqueous_explosive_volcanism/links/02e7e515b982d63bc3000000.pdf. Accessed December 15, 2016.

Zimanowski, B., R. Büttner, P. Dellino, J.D.L. White, and K.H. Wohletz. 2015. Magma-water interaction and phreatomagmatic fragmentation. Pp. 473-484 in Encyclopedia of Volcanoes, 2nd Ed., H. Sigurdsson, B. Houghton, S.R. McNutt, H. Rymer, and J. Stix, eds. San Diego, CA: Academic Press. doi:10.1016/B978-0-12-385938-9.00026-2.

Appendix A

Volcano Databases

Descriptions of Volcanoes and Eruption Histories:

• The Smithsonian Global Volcano Program (Volcanoes of the World) has reports, eruption history, deformation episodes, and general geologic information for Holocene active volcanoes. Satellite-based SO_2 emission data are now included. See http://volcano.si.edu/search_volcano.cfm

• The National Oceanic and Atmospheric Administration has a database of significant volcanic eruptions with Volcano Explosivity Index (VEI), number of fatalities, and damage estimates, as well as a database of volcanic ash advisories. See https://www.ngdc.noaa.gov/hazard/volcano.shtml.

• LaMEVE (Large Magnitude Explosive Volcanic Eruptions) has information on Quaternary active volcanoes, including volcano type and eruptive history if known for VEI ≥4 eruptions. This database is part of VOGRIPA (Volcanic Global Risk Identification and Analysis Project). See http://www.bgs.ac.uk/vogripa/index.cfm.

Unrest Prior to Eruptions:

• WOVOdat is a database of volcanic unrest, including instrumentally and visually recorded changes in seismicity, ground deformation, gas emissions, and other parameters from observatories' normal baselines. See http://www.wovodat.org.

Geophysical Data:

• IRIS (Incorporated Research Institutions for Seismology) provides seismic data through http://ds.iris.edu/ds/nodes/dmc, and UNAVCO provides geodetic data through https://www.unavco.org/data/data.html.

• The COMET (Centre for Observation and Modelling of Earthquakes, Volcanoes, and Techtonics) Volcano Deformation Database includes data over volcanoes from the European Space Agency's Sentinel satellites. See http://volcanodeformation.blogs.ilrt.org.

• Volcano Deformation Database, part of the Global Volcano Model (GVM). See http://globalvolcanomodel.org/gvm-task-forces/volcano-deformation-database.

Geochemical, Gas, and Thermal Data:

• IEDA (Interdisciplinary Earth System Alliance)/EarthChem provides geochemical information, including some volcanic gas data. The data are not linked to eruption data. See http://www.earthchem.org.

• The GeoRoc (Geochemistry of Rocks of the Oceans and Continents) database contains geochemical information. See http://georoc.mpch-mainz.gwdg.de/georoc.

• The Multi-Satellite Volcanic Sulfur Dioxide L4 Long-Term Global Database V2 provides data on volcanic SO_2 emissions derived from ultraviolet satellite measurements since October 1978. See ftp://measures.gsfc.nasa.gov/data/s4pa/SO2/MSVOLSO2L4.2.

- MAGA (MApping GAs emissions) is a volcanic gas database focused on the Mediterranean region. See http://www.magadb.net.
- The ASTER (Advanced Spaceborne Thermal Emission and Reflection Radiometer) Volcano Archive provides thermal and optical imagery for all of the world's volcanoes. See http://ava.jpl.nasa.gov.
- The MODVOLC algorithm provides a tool to find near-real-time thermal monitoring data from MODIS (Moderate Resolution Imaging Spectroradiometer) sensors. See http://modis.higp.hawaii.edu/cgi-bin/modisnew.cgi.

Appendix B

Workshop Participants

Tim Ahern, Incorporated Research Institutions for Seismology
Diaa Ahmed, Independent Research Professional
Kyle Anderson, U.S. Geological Survey
Ben Andrews, Smithsonian Institution
Gerald Bawden, National Aeronautics and Space Administration
Bruce Beaudoin, New Mexico Institute of Mining and Technology
Ben Black, City University of New York
Costanza Bonadonna, University of Geneva
Mike Burton, University of Manchester
Simon Carn, Michigan Technological University
Raymond Cas, Monash University and University of Tasmania
Katharine Cashman, University of Bristol
William Chadwick, Oregon State University and Pacific Marine Environmental Laboratory
Remy Chappetta, National Academies of Sciences, Engineering, and Medicine
Amy Chen, National Aeronautics and Space Administration
Amanda Clarke, Arizona State University
Charles Connor, University of South Florida
Michelle Coombs, U.S. Geological Survey
Kari Cooper, University of California, Davis
Fidel Costa Rodriguez, Nanyang Technological University
Leah Courtland, University of Indianapolis
Donald Dingwell, University of Munich
Josef Dufek, Georgia Institute of Technology
Eric Dunham, Stanford University
Eric Edkin, National Academies of Sciences, Engineering, and Medicine
Elizabeth Eide, National Academies of Sciences, Engineering, and Medicine
Sonia Esperança, National Science Foundation
Kristen Fauria, University of California, Berkeley
Tobias Fischer, University of New Mexico
Carol Frost, National Science Foundation
James Gardner, The University of Texas at Austin
Dennis Geist, National Science Foundation
Kimberly Genareau, University of Alabama
Thomas Giachetti, University of Oregon
Courtney Gibbs, National Academies of Sciences, Engineering, and Medicine
Guido Giordano, Università Roma Tre
Helge Gonnermann, Rice University
Julia Hammer, University of Hawaii, Manoa
Bruce Houghton, University of Hawaii, Manoa
Chris Huber, Brown University
Steve Ingebritsen, U.S. Geological Survey
Claude Jaupart, University of Paris
Jeffrey Johnson, Boise State University
Gill Jolly, GNS Science
Meghan Jones, Massachusetts Institute of Technology and Woods Hole Oceanographic Institution
Leif Karlstrom, University of Oregon

Kerry Key, Scripps Institution of Oceanography
Nickolay Krotkov, National Aeronautics and Space Administration
John LaBrecque, Global Geodetic Observing System
Peter LaFemina, The Pennsylvania State University
Anne Linn, National Academies of Sciences, Engineering, and Medicine
Paul Lundgren, Jet Propulsion Laboratory
Charles Mandeville, U.S. Geological Survey
Michael Manga, University of California, Berkeley
Erin Markovich, National Academies of Sciences, Engineering, and Medicine
Warner Marzocchi, National Institute of Geophysics and Volcanology
Larry G. Mastin, U.S. Geological Survey
Robin Matoza, University of California, Santa Barbara
Glen Mattioli, UNAVCO, Inc.
Gari Mayberry, U.S. Agency for International Development
Brendan McGovern, National Academies of Sciences, Engineering, and Medicine
Steve McNutt, University of South Florida
Calvin Miller, Vanderbilt University
Tom Murray, U.S. Geological Survey
Augusto Neri, National Institute of Geophysics and Volcanology
Sarah Ogburn, U.S. Geological Survey
Michael Pavolonis, National Oceanic and Atmospheric Administration
Benjamin Phillips, National Aeronautics and Space Administration
Terry Plank, Columbia University
Michael Poland, U.S. Geological Survey
Matthew Pritchard, Cornell University
Michael Ramsey, University of Pittsburgh
Nicholas Rogers, National Academies of Sciences, Engineering, and Medicine
Diana Roman, Carnegie Institution for Science
Brandon Schmandt, University of New Mexico
Paul Segall, Stanford University
Freysteinn Sigmundsson, University of Iceland
Thomas Sisson, U.S. Geological Survey
S. Adam Soule, Woods Hole Oceanographic Institution
Steve Sparks, Bristol University
Greg Valentine, University of Buffalo
Jennifer Wade, National Science Foundation
Greg Waite, Michigan Technological University
Paul Wallace, University of Oregon
Dorsey Wanless, Boise State University
Peter Webley, University of Alaska, Fairbanks
Aleeza Wilkins, U.S. Geological Survey
Heather Wright, U.S. Geological Survey

Appendix C

Biographical Sketches of Committee Members

Michael Manga (*Chair*) is a professor in the Department of Earth and Planetary Science at the University of California, Berkeley. His research focuses on processes involving fluids in natural systems, including problems in physical volcanology, geodynamics, and hydrogeology using combinations of theoretical, numerical, and experimental approaches and field observations. Dr. Manga has served on advisory committees, including the National Research Council (NRC) Committee on New Research Opportunities in the Earth Sciences at the National Science Foundation (NSF), the Kavli Institute for Theoretical Physics, and Physics Today. He was named a MacArthur Foundation Fellow in 2005, and is a fellow of the Geological Society of America (GSA) and the American Geophysical Union (AGU). He is also a recipient of several awards, including the European Geoscience Union's Bunsen Medal for distinguished research in geochemistry, mineralogy, petrology, and volcanology; and GSA's Donath Medal and AGU's James B. Macelwane Medal, both for significant contributions by an outstanding early career scientist. Dr. Manga received a B.Sc. in geophysics from McGill University, and an M.S. in engineering sciences and a Ph.D. in earth and planetary sciences from Harvard University.

Simon A. Carn is an associate professor in the Department of Geological and Mining Engineering and Sciences at Michigan Technological University. His research focuses on the application of remote sensing data to studies of volcanic degassing, volcanic eruption clouds, and anthropogenic pollution, with a particular emphasis on SO_2, which plays an important role in climate. Dr. Carn has participated in several advisory activities, including a technical advisory committee for a United Nations Project on volcano risk reduction in Goma, Democratic Republic of the Congo, and as secretary of the International Association of Volcanology and Chemistry of the Earth's Interior (IAVCEI) Remote Sensing Commission. His participation on science teams for satellite measurements of sulfur dioxide, ozone, air quality, and climate earned him the William T. Pecora Award (National Aeronautics and Space Administration [NASA]/U.S. Department of the Interior) and the NASA Group Achievement Award. Dr. Carn received a B.A. in natural science, geology, from Exeter College, University of Oxford, United Kingdom; a D.E.A. in volcanology and magmatic processes from Université Blaise Pascal, France; and a Ph.D. in volcanology from St. Catharine's College, University of Cambridge, United Kingdom.

Katharine V. Cashman is the AXA Endowed Chair of Volcanology at the University of Bristol. Before being recruited to Bristol in 2011, she spent 20 years on the faculty of the University of Oregon. One of the top volcanologists in the world, Dr. Cashman's research focuses on the evolution of magma within the Earth's crust and how its path to the surface triggers volcanic eruptions. She uses a combination of field volcanology, igneous petrology, kinetics, microscopy, and fluid dynamics to address the fundamental problem of how volcanoes work. Dr. Cashman is a former member of the scientific advisory committee for the Soufrière Hills Volcano, Montserrat, and a former president of AGU's Volcanology, Geochemistry, and Petrology Section. She is a recipient of AGU's Bowen Award for her outstanding contributions to volcanology. Dr. Cashman is a fellow of the AGU, the American Academy of Arts and Sciences, and the Royal Society. She is a member of the Academia Europaea and the National Academy of Sciences. She received a B.A. in geology/biology from Middlebury College, an M.S. from Victoria University, New Zealand, and a Ph.D. in geological sciences from Johns Hopkins University.

Amanda B. Clarke is an associate professor in the School of Earth and Space Exploration at Arizona State University, and an associated researcher at the Istituto Nazionale di Geofisica e Vulcanologia in Pisa, Italy. Her research focuses on the physics of explosive volcanic eruptions, applying fluid mechanics, laboratory experiments, and numerical methods to fundamental questions in volcanology. She has also been involved in volcano monitoring, working at the Montserrat Volcano Observatory during a long sequence of devastating explosive eruptions in 1997, and in assessing eruption impacts, studying the economic and cultural effects of a 1993 eruption of Mayon Volcano on four nearby villages. Dr. Clarke is a recipient of the Wager Medal, awarded by IAVCEI for her contributions to volcanology. She currently chairs IAVCEI's Commission on Explosive Volcanism. She received a B.S. in aerospace engineering and a B.A. in philosophy from the University of Notre Dame and a Ph.D. in geosciences from The Pennsylvania State University.

Charles B. Connor is a professor in the School of Geosciences at the University of South Florida. His primary research interests are in geophysics and volcanology, particularly the development of volcanic hazard and risk models, and modeling heat and mass transfer in volcanoes to improve volcano monitoring. He has studied volcanoes around the world, using a combination of field work, laboratory studies of geologic processes, and numerical simulation to understand the basic physical controls on these processes. Dr. Connor is a co-founder of the IAVCEI Commission on Statistics in Volcanology, a former member of the U.S. Department of Energy panel to assess volcanic hazards at Yucca Mountain, and a former member of the NRC Committee to Review the Volcano Hazards Program of the U.S. Geological Survey. He received a B.S. in geology and a B.A. in anthropology from the University of Illinois at Urbana-Champaign, and an M.S. and a Ph.D. in geology from Dartmouth College.

Kari M. Cooper is a professor in the Department of Earth and Planetary Sciences at the University of California, Davis. Her research is aimed at understanding how magmas evolve and interact with each other and with their surroundings, with a particular focus on dating volcanic crystals to understand the time scales and conditions of magma storage and mobilization prior to eruptions. Dr. Cooper has presented keynote and invited talks and co-convened several special sessions on these topics, most recently on comparing crystallization histories with eruption histories. She is a fellow of the GSA. Dr. Cooper received a B.A. in geology from Carleton College, an M.S. in geology from the University of Washington, and a Ph.D. in geochemistry from the University of California, Los Angeles.

Tobias Fischer is a professor in the Department of Earth and Planetary Sciences at the University of New Mexico. His research focuses on the geochemistry of gases discharging from active volcanoes to help constrain the processes that lead to explosive eruptions and to investigate eruption precursors. He maintains a laboratory for gas analyses and the development of field instrumentation for these purposes. Dr. Fischer chairs the Deep Carbon Degassing project, an international initiative aimed at better constraining carbon degassing from the Earth's interior, and is a member of the steering committee for the Deep Carbon Observatory's reservoirs and fluxes program. He received a Vordiplom from Albert-Ludwigs Universiät, Freiburg, Germany, and an M.S. and a Ph.D. from Arizona State University, all in geology.

Bruce Houghton is the Gordon A. MacDonald Professor of Volcanology at the University of Hawaii at Manoa, and the Hawaiian State Volcanologist. He is also the science director for the Federal Emergency Management Agency–funded National Disaster Preparedness Training Center at the University of Hawaii. Dr. Houghton's research focuses on understanding the mechanisms of explosive eruptions by constraining the nature of the eruptions and their products in near real time. His natural hazards research examines knowledge, perceptions, and preparedness for volcanic eruptions, tsunamis, flooding, and sea-level rise. Dr. Houghton has served on numerous committees focused on different aspects of volcanism, and is currently an executive member of IAVCEI commissions on tephra hazard modeling and on cities on volcanoes. He is a fellow of the GSA, and a former president of the Geological Society of New Zealand. He is a fellow of the Royal Society of New Zealand. He received a B.Sc. in geology from the University of Auckland, and a Ph.D. in volcanology from the University of Otago, New Zealand.

Jeffrey B. Johnson is an associate professor of geophysics in the Department of Geosciences at Boise State University. Before coming to Boise State in 2012, he was a research professor at the New Mexico Institute of Mining and Technology. Dr. Johnson's research interests focus on multidisciplinary geophysical study of eruptive processes, infrasound science and sensor development, and volcano monitoring. He recently organized an international workshop on interdisciplinary monitoring and integration of new technologies for a volcano in Santiaguito, Guatemala. He is a founding member of the IAVCEI/IASPEI (International Association of Seismology and Physics of the Earth's Interior) committee for volcano acoustics, and a member of the volcano seismology commission. Dr. Johnson received a B.S. in geological and environmental sciences and an M.S. in geophysics from Stanford, and a Ph.D. in geophysics from the University of Washington.

Terry A. Plank is the Arthur D. Storke Memorial Professor in the Department of Earth and Environmental Sciences at the Lamont Doherty Earth Observatory of Columbia University. She uses a variety of geochemical tracers to study the origin and evolution of magmas and to determine the concentration of water in magma, which drives both melt formation and explosive eruptions. Dr. Plank has served on several recent advisory committees, including the executive committee for the Deep Carbon Observatory, and the National Academies U.S. National Committee for Geodesy and Geophysics. She received the European Association for Geochemistry's Houtermans Medal and the GSA's Donath Medal for outstanding contributions from an early career scientist, and was named a MacArthur Foundation Fellow in 2012. Dr. Plank is a fellow of the AGU, the GSA, the Geochemical Society, and the Mineralogical Society of America. She is a member of the National Academy of Sciences. She received a B.A. in earth sciences from Dartmouth College and a Ph.D. in geosciences from Columbia University.

Diana C. Roman is a staff scientist in the Department for Terrestrial Magnetism at the Carnegie Institution for Science. Before joining the staff at Carnegie, she spent 5 years on the faculty of the University of South Florida. Dr. Roman's research straddles the boundary between volcanology and seismology, and focuses on understanding source processes of volcanic earthquakes, volcano-fault interaction, and the structure and dynamics of magma transport and storage systems. Over the past few years, she has co-convened several special sessions on these topics, including one on outstanding challenges in the seismological study of volcanic processes (AGU, 2014). Dr. Roman is a recipient of IAVCEI's George Walker Award, which recognizes achievements of a recent outstanding graduate in volcanology. She received a B.S. in applied economics from Cornell University, and an M.S. and a Ph.D. in geological sciences from the University of Oregon.

Paul Segall is a professor in the Department of Geophysics at Stanford University. He studies volcanic processes by measuring surface deformation, determining the geometry of magma chambers and how they change over time, and developing models to understand the physics of magma migration leading to volcanic eruptions. He uses similar approaches to study active faulting. Dr. Segall is a former chair of the Plate Boundary Observatory Steering Committee, member of the UNAVCO Board of Directors, and member of the National Academies Committee to review the Volcano Hazards Program of the U.S. Geological Survey. He is a fellow of the AGU and the GSA, and a recipient of the AGU's Macelwane Medal (early career award) and Whitten Medal for outstanding achievement in research on the form and dynamics of the Earth and planets. He is a member of the National Academy of Sciences. He received a B.A. and an M.S. in earth sciences from Case Western Reserve University and a Ph.D. in geology from Stanford University.

Appendix D

Acronyms and Abbreviations

AIRS Atmospheric Infrared Sounder
ALOS Advanced Land Observing Satellite
ASCENDS Active Sensing of CO_2 Emissions over Nights, Days, & Seasons
ASTER Advanced Spaceborne Thermal Emission and Reflection Radiometer
AVHRR Advanced Very High Resolution Radiometer
AVO Alaska Volcano Observatory

CALIPSO Cloud-Aerosol Lidar and Infrared Pathfinder Satellite Observation
CIDER Cooperative Institute for Dynamic Earth Research
CIG Computational Infrastructure for Geodynamics
CMG Collaboration in Mathematical Geosciences
COSMO-SkyMed Constellation of Small Satellites for Mediterranean basin Observation
CSEDI Cooperative Studies of the Earth's Deep Interior

DOAS Differential Optical Absorption Spectroscopy

ENSO El Niño–Southern Oscillation

FTIR Fourier Transform Infrared Spectroscopy

GeoPRISMs Geodynamic Processes at Rifting and Subducting Margins
GNSS Global Navigation Satellite System
GOES Geostationary Operational Environmental Satellite
GPS Global Positioning System

IASI Infrared Atmospheric Sounding Interferometer
InSAR interferometric synthetic aperture radar
IR infrared
IRIS Incorporated Research Institutions for Seismology

MISR Multi-angle Imaging SpectroRadiometer
MLS Microwave Limb Sounder
MODIS Moderate Resolution Imaging Spectroradiometer

NASA National Aeronautics and Space Administration
NCAR National Center for Atmospheric Research
NERSC National Energy Research Scientific Computing Center

NSF	National Science Foundation
OMI	Ozone Monitoring Instrument
OMPS	Ozone Mapping and Profiler Suite
PASSCAL	Portable Array Seismic Studies of the Continental Lithosphere
PDC	pyroclastic density current
PREVENTS	Prediction of and Resilience against Extreme Events
SAGE	Stratospheric Aerosol and Gas Experiment
SEES	Science, Engineering, and Education for Sustainability
TanDEM-X	TerraSAR-X add-on for Digital Elevation Measurement
USGS	U.S. Geological Survey
VDAP	Volcano Disaster Assistance Program
VEI	Volcano Explosivity Index
XSEDE	Extreme Science and Engineering Discovery Environment